Authentic and Inauthentic Places in Tourism

With the rise of post-truth and fake news, a thorough examination of authenticity has never been so relevant. This book explores the geography of authenticity, investigating a wide variety of places used by tourists. Not only does it assess what might be described as the more traditional objects for examination – places such as the city, the countryside and the coast – it also includes chapters on art and place, hipster places, gentrification, heritage sites, film locations, photographed places and eventful places.

Using a wide-angled lens on places reveals linkages and possibilities, enabling the book to skate across the surface of the geography of authenticity, locating the magically real heritage site, the poignant replica, the authenticated theme park, the unmasked carnival. In focusing on authentic and inauthentic places, this text provides a useful contribution to the understanding of how places are changing, how they are perceived and how authenticity is embodied and performed within them. *Authentic and Inauthentic Places in Tourism* is an insightful study and an essential read for those involved in the study of geography, tourism, urban studies, culture and heritage.

Jane Lovell is a Senior Lecturer at Canterbury Christ Church University. Jane's PhD thesis examined staged authenticity in historic cities. She is an authenticist (and is very real). She is Chair of Canterbury Heritage Partnership, an Associate Fellow of UCL Institute of the Americas and member of the Council of the British Association of Canadian Studies.

Chris Bull is the former Head of the Department of Sport Science, Tourism and Leisure, Canterbury Christ Church University. He is the co-author of several texts: *The Geography of Rural Resources* (1984); *An Introduction to Leisure Studies* (2003); and *Sports Tourism: Participants, Policy and Providers* (2004 and 2009).

Contemporary Geographies of Leisure, Tourism and Mobility

Series Editor: C. Michael Hall

Professor at the Department of Management, College of Business and Economics, University of Canterbury, Christchurch, New Zealand

The aim of this series is to explore and communicate the intersections and relationships between leisure, tourism and human mobility within the social sciences.

It will incorporate both traditional and new perspectives on leisure and tourism from contemporary geography – for example, notions of identity, representation and culture – while also providing for perspectives from cognate areas such as anthropology, cultural studies, gastronomy and food studies, marketing, policy studies and political economy, regional and urban planning, and sociology, within the development of an integrated field of leisure and tourism studies.

Also, increasingly, tourism and leisure are regarded as steps in a continuum of human mobility. Inclusion of mobility in the series offers the prospect to examine the relationship between tourism and migration, the sojourner, educational travel and second home and retirement travel phenomena.

For a full list of titles in this series, please visit www.routledge.com/series/SE0522

The series comprises two strands:

Contemporary Geographies of Leisure, Tourism and Mobility aims to address the needs of students and academics, and the titles will be published in hardback and paperback. Titles include:

Tourism and Climate Change
Impacts, adaptation and mitigation
C. Michael Hall, Stefan Gössling and Daniel Scott

Tourism and Citizenship
Raoul V. Bianchi and Marcus L. Stephenson

Routledge Studies in Contemporary Geographies of Leisure, Tourism and Mobility is a forum for innovative new research intended for research students and academics, and the titles will be available in hardback only. Titles include:

International Tourism Development and the Gulf Cooperation Council States
Challenges and Opportunities
Edited by Marcus L. Stephenson and Ala Al-Hamarneh

Authentic and Inauthentic Places in Tourism
From Heritage Sites to Theme Parks
Jane Lovell and Chris Bull

Authentic and Inauthentic Places in Tourism

From Heritage Sites to Theme Parks

Jane Lovell and Chris Bull

Routledge
Taylor & Francis Group
LONDON AND NEW YORK

First published 2018
by Routledge
2 Park Square, Milton Park, Abingdon, Oxon OX14 4RN

and by Routledge
711 Third Avenue, New York, NY 10017

Routledge is an imprint of the Taylor & Francis Group, an informa business

British Library Cataloguing-in-Publication Data
A catalogue record for this book is available from the British Library

Library of Congress Cataloging-in-Publication Data
A catalog record for this book has been requested

ISBN: 978-1-138-93670-6 (hbk)
ISBN: 978-1-315-67668-5 (ebk)

Typeset in Times New Roman
by Cenveo Publisher Services

Contents

Illustrations

Figures

Tables

1 Introduction

Places and authenticity

This book explores the geography of authenticity, investigating a wide variety of places used by tourists, ranging from heritage sites, wilderness and seaside environments to theme parks, ruins and gentrified cityscapes, locating the fantastical heritage site, the poignant replica, the authenticated theme park, the unmasked event, the detail within the photograph and the mighty-real castle. This book identifies how, as globalisation, gentrification, digitisation and virtual realities continue to evolve, authenticity is increasingly germane to placemaking and acts of authentication, both tangible and intangible, occur on granular and meta scales in a variety of locations. As Lovell (2013) discovered when researching staged authenticity, it is an international concept, recognised by tourists from different countries, spanning all continents. The strength of this text is its breadth; using a wide-angled lens to explore place authenticity reveals linkages and possibilities across different disciplines which make the work significant.

Revisiting the topic of authenticity has never been so relevant. The term 'post-authenticity' (Labadi, 2010, p. 78) has been used to describe the inseparable mix of fake and real styles in the built heritage environment and postmodern theories suggest that we are beyond separating the real from the copy. With the rise of post-truth, 'authentic' politicians, fake news and the concept that we live in a 'fake world' of hypernormalisation and spectacle there has been what amounts to a popular 'authenticity turn' or, more accurately, an 'inauthenticity turn'. In the climate of post-postmodernism, where virtual combinations of fiction and fact seem unfilterable, the levelling effect of digitisation ensures that verification is obscured and memes are mass-reproduced, we are increasingly unsure about our reality. However, truth is not simply unvarnished or 'post' and as the book demonstrates, there are many variations in between.

Tourism does not occur within a vacuum; some type of place, space or landscape is required, whether as the specific location of experiences, an incidental backdrop, or as the site of essential resources upon which certain activities depend. As the authors of this book have noted, whenever the (in)authenticity of places is mentioned, the default response is to cite Las Vegas. However, places do not exist on a fake or real continuum where they are compared to Las Vegas and Disney World; instead, mobility enables discerning people to compare the aesthetics of a wide range of types of different places (Maitland and Smith,

2009). The degree to which the places that tourists visit are perceived as authentic or inauthentic and the extent to which places are modified, artificially constructed and organised or marketed as 'real' are essential ingredients of both the tourist and local experience and the places themselves.

Many texts deal with tourism places, not least because urban theorists and geographers have embraced tourism as a legitimate area for their attention. These include books on specific tourist places such as towns and cities (Law, 1993, 1996; Page, 1995), the countryside (Butler *et al.*, 1998; Sharpley and Sharpley, 1997, 2007), the seaside (Shaw and Williams, 1997), islands (Lockhart and Drakakis-Smith, 1997) and border regions (Timothy, 2001), as well as general tourism geographies (Aitchison *et al.*, 2000; Hall and Page, 2006; Lew *et al.*, 2008; Pearce, 1995; Robinson, 1976; Shaw and Williams, 2002, 2004; Williams, 1998); but few of these address the issue of authenticity as a central theme. Spatial theorists have also encompassed authenticity within the context of the creativity and practice of space, while not stating it directly, and others describe placelessness or nostalgic places (Augé, 1995; Lefebvre, 1991; Massey, 2005; Relph, 1976 [2008]; Soja, 1999; Tuan, 1977).

There has been an increasing interest in the theorisation of authenticity in tourism studies (Cohen, 1979a, 1988a, 2007; Cohen and Cohen, 2012; Jamal and Hill, 2002; Kim and Jamal, 2007; Hollinshead, 1998; Kirshenblatt-Gimblett, 1998; Lovell, 2013; Ram *et al.*, 2015; Reisinger and Steiner, 2005; Wang, 1999; Bruner, 1994; Moscardo and Pearce, 1986) and its relevance has also been addressed as a marketing concept (Pine and Gilmore, 1999; York, 2016). Authenticity has been linked to philosophy, sociology, psychology, anthropology and cultural studies; ontological approaches include critical realism, research approaches encompassing ethnography (Bruner, 1994), empiricism (Littrell *et al.*, 1993) and phenomenology (Hayller and Griffin, 2005; Pernecky and Jamal, 2010; Lovell, 2013). Authenticity is a contested concept, rife with misdirection and contradictions, and has been described as an inchoate term, or a binary, schematic concept, debated in the light of postmodernism to the extent of questioning if it is ultimately a 'red herring', purely designed by and for theorists (Bruner, 1991, p. 241), a 'quixotic quest' (Crang, 1996, p. 415), an obsolete notion (Reisinger and Steiner, 2006) or performed as authentication (Cohen and Cohen, 2012). It is important to note that the focus of much of the relevant literature has not been explicitly directed at places. It is also worth noting that some authors have divided authenticity into studies of objects (in other words, places) and experiences (Wang, 1999; Jamal and Hill, 2002; MacDonald, 2009; Steiner and Reisinger, 2006). This work suggests that there is a stronger, intersubjective dialogue between the two, blending experience and place, suggesting that the pursuit of authenticity involves intertextuality and mixed realities.

Aura

Early authenticity theory is evident through the works of Greek philosophers, such as Plato's *Theory of Forms*. In the allegory of the cave, the shadows seen within by the light of the flickering fire are revealed to be illusory when a person steps out into the sun. This was intended to illustrate the effect of education. The forms were ideals which could never be realised in the material world. This relates to one of the key influences on this book, the work of the philosopher Walter Benjamin (1936 [2008]), who addresses the concept of aura in his key text *The Work of Art in the Age of Mechanical Reproduction*. Benjamin (ibid., p. 39) described aura as having 'cultic value' or 'a unique manifestation of a remoteness', 'no matter how near it may be'. What makes an auratic object singular is its unapproachability, distinctiveness, rarity and, ultimately, its distance. Benjamin contended that people try to bring the auratic closer through mass-reproduction techniques and, in doing so, the idea of 'genuineness' begins to 'wobble' and aura 'shrinks' and 'fades'. Benjamin (ibid., p. 40) goes on to suggest that when objects are reproduced, aura becomes increasingly vague: 'The more the cultic value of the image is secularised, the less specific ideas of the substratum of its uniqueness become.' His contention is that when objects are mass-reproduced, aura is *replaced* by authenticity which 'supplants cultic value'. Benjamin argues that people try to demystify objects by explaining what makes things genuine through empirical methods of authentication. The central paradox of a tourist search for authenticity is that the closer we try to bring iconic places, the more remote they can seem.

Space and place

The book focuses on places and authenticity, but the idea of place could be said to be a construct. Many spatial theorists debate the notion of which came first, place or space, which is tantamount to asking whether places can be inherently authentic, or whether authenticity is a notion projected by people. Cresswell (2006, p. 33) describes place as a 'construction of humanity, but a necessary one', meaning that place is fundamental to the construction of human society in a way that many other social constructions are not. This suggests that place is a fundamental, authentic part of social life, rather than the two notions acting as 'mutually constitutive'. Cohen and Cohen observe, referring to Smith (1987, p. 28, cited in Cohen and Cohen, 2012, pp. 1301–2) that 'people evoke places through practice – "human beings are not placed, they bring place into being"'. Crouch (2000, p. 95) suggests that tourist experiences in destinations are 'creative, active and purposive' and practised, for example, walking as an act of authentic placemaking (Pink *et al.*, 2010, p. 6). In the relationship of the social to the geographical, there is an argument that the geographical shapes the social, as Malpas (1999, cited in Cresswell, 2006, p. 30) argues: structure and meaning is created by place, which societies then design themselves around. Space is frequently portrayed as fluid – for example, in the work of Massey (2005) and Kwon (2004), who suggest that places are always in motion, fluctuating and assembling themselves and the 'circulation' (Coleman and Crang, 2002; Collins-Kreiner, 2010; Jansson, 2010) of places underpins their malleability. De Certeau (1984, cited in Cresswell, 2006, p. 39) posits that 'place is an empty grid, space is created by practice', so place is an idea, space is lived.

Several spatial thinkers have discussed theories where authenticity is implicit but not directly addressed. A triad of conceptual spaces developed by Lefebvre (1991) and Soja (1999) explores the different aspects of space: 'First Space' is physical and objective, concerning the built environment; 'Second Space' is intellectual and constructed, occupied by planners, policy-makers and architects; 'Third Space' is practised and lived, incorporating movement, smell, sound. These spaces were conceived by Lefebvre as approximations which may intermingle and shift. The spaces loosely align with three types of authenticity – objective, constructive and experiential – defined in a seminal work by Wang (1999), which will be discussed throughout this introduction. The anthropologist Augé (1995) developed the theory of non-places as transitory locations with mechanical processing interfaces, such as airports. These suggest a processual, cool form of non-authenticity, rather than inauthenticity, although airports can be the scene of highly charged emotional experiences. Lastly, in his influential study of *Place and Placelessness*, Relph (1976 [2008], pp. 51–4) identified various states of insideness and outsideness, including existential outsideness, which describes alienation and the 'non-place'.

Theoretical perspectives on tourism and authenticity

The study of authenticity (like the search for aura by tourists) can be pursued, but the closer we get to understanding it, the more distant it may seem. According to Wearing *et al.* (2010, p. 27) 'the notion that tourism is a search for authenticity is one of the most well-known and well-established theoretical debates in the study of tourism'. Interest in authenticity in tourism studies is generally regarded as dating from the work of Boorstin (1961 [1964]), who regarded mass tourism as a homogeneous tourism product involving a commodification of culture, producing contrived tourist experiences constructed around 'pseudo-events'. His somewhat elitist view suggested that tourists seldom liked the authentic, preferring instead approximations of the 'real' as well as the comfort and security of the protective 'environmental bubble' that insulates them from the strangeness of the host environment and local people.

In contrast, MacCannell (1973), writing a decade later, offered a very different perspective, suggesting that, rather than embracing the inauthentic, tourists were instead motivated to travel in search of authenticity in other places and cultures. Such desires were driven by a need to escape the humdrum reality of their everyday lives and experience something different. Thus, Cohen (1988a, p. 374), paraphrasing MacCannell, suggests that:

> the alienated modern tourist in quest of authenticity hence looks for the pristine, the primitive, the natural, that which is as yet untouched by modernity… and hopes to find it in other times and other places… since it is absent from his own world.

However, MacCannell believed that such needs were destined to fail due to the organisation and priorities of the tourism industry. The real world of host communities that tourists would ideally wish to encounter would involve an obvious intrusion into local people's lives that would be unacceptable. According to MacCannell's thesis, 'the people being observed and local tourism entrepreneurs gradually come to construct back stages in a contrived and artificial manner' (Urry, 2002, p. 9) and, as a result, tourists are only able to experience, what MacCannell (1973, p. 94) refers to as 'staged authenticity'. Authenticity is implied here to be a form of managed commodification, a way for local people to retain control over their image: 'The term "tourist" is increasingly used as a derisive label for someone who seems content with his obviously inauthentic experiences' (ibid.). Although tourists wish to experience the real lives of others in what might be referred to as authentic, 'back' regions, they are usually unable to travel beyond the 'front' regions (Wearing *et al.*, 2010) while searching for them. Thus, according to Urry (2002, p. 7), MacCannell argues 'that "pseudo-events" result from the social relations of tourism and not from an individualistic search for the inauthentic'.

While the respective views of Boorstin and MacCannell provoke interesting debate, subsequent work has highlighted the shortcomings of their claims. Both treat tourists as a homogeneous group, whereas various writers have highlighted the fact that there are different types of tourists with different demands, motivations and consumption patterns (Cohen, 1979a, 1988a; Pearce, 1995). The other problem is that authenticity is used in a rather simplistic, uncritical way, while the reality is that it is a multifaceted, complex concept.

To provide a conceptual clarification, Wang (1999), for example, considers three separate approaches to authenticity: objective, constructive and existential. Cohen (2007) contends that 'constructed' authenticity, the third category in Wang's typology, is not a distinct discourse but should rather be conceived as a hybrid of the other two, and goes on to reiterate this in Cohen and Cohen (2012, p. 1296), arguing that Wang's three discourses are not on the same level. While objective (object) authenticity and existential (subjective) authenticity denote different types of (personally experienced) authenticity, constructed authenticity does not; rather it relates to the process of social construction of the other two types. As stated in the introduction to this chapter, while Wang has separated these different types of authenticity, there is a strong relationship between them; places are an 'assemblage' (MacDonald, 2009, p. 4; Harrison, 2013, p. 33) or composite experience: 'The types of space and the complex interactions between them should be studied together as fundamental and intertwined knowledge sources, for this is what "being-in-the-world is all about"' (Soja, 1999, p. 262).

Staged authenticity

As detailed earlier, MacCannell's contention is that tourists who search for depth of authenticity can also be a dupe of staged authenticity which is 'invented,' manufactured, or self-consciously provided as a product for tourist consumption, by hosts (Bruner, 1991, p. 241; MacCannell, 1973, p. 106). MacCannell was the

Table 1.1 MacCannell's stages, as summarised by Moscardo and Pearce

Stage	*Description*
1	Goffman's first region, which tourists supposedly seek to penetrate and overcome
2	Here MacCannell postulates the existence of a tourist front region that has been decorated to superficially resemble a back region
3	A front region totally organised to represent a back region
4	Environments open to outsiders but which should be essentially considered a back region
5	Environments with more limited access to outsiders
6	Goffman's 'back region'

Sources: MacCannell, 1973, 1976; Pearce and Moscardo, 1986, p. 125.

first theorist to outline a six-point categorisation of staged authenticity in his discussion of staged authenticity, using Goffman's (1959) front and back regions (see Table 1.1). Goffman had first coined the terms 'front' and 'back' regions (pp. 92–9) to demonstrate the realms of social performance. The front region is for the host and guest and the back region is for the hosts to 'relax and prepare'. There is also an overtly staged backstage area where people can seemingly see behind the scenes, a 'living museum'. Moscardo and Pearce (1986, p. 125) then used the MacCannell typologies to produce nine 'classifications of tourist experiences' using encounters with people in the different regions, although their approach was contested by Turner and Manning (1986, p. 138), who suggested that the regions were interconnected.

A further perspective that is germane to the contents of this text is contained in Cohen's (1979a) 'impressions of scene'. This idea is outlined in the 'four square model of tourist situations' in Figure 1.1, based on a Latin square or 'four cell model' (Cohen, 1979a, p. 26, in Pearce and Moscardo, 1986, p. 125).

Here the first box, 'real real', describes a scenario where the object or site is authentic and is recognised as such; 'staged real' (box B) is where the space is artificial/contrived but where the visitor fails to recognise this; 'real staged' (C) is where there is a suspicion of staging and where authenticity is questioned; and the final box (D) 'staged staged' is where the tourist space is accepted as contrived and inauthentic.

A variation of staging was employed by Lovell (2013) while researching staged authenticity in 15 English historic towns and cities. This suggests that concepts such as authentic and inauthentic, real and fake, do not exist in a

Figure 1.1 Four cell model of tourist situations

Nature of scene	Real	Staged
Real	A authentic and recognised as such	C suspicion of staging, authenticity questioned
Staged	B Failure to recognise contrived tourist space	D Recognised contrived tourist space

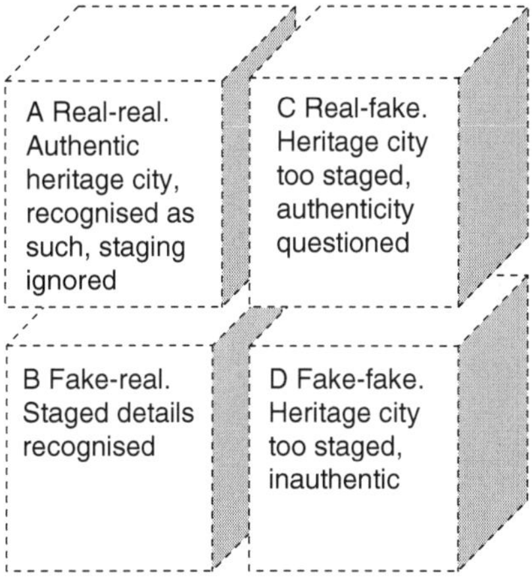

Figure 1.2 Impressions of staging in historic cities

Sources: Lovell, 2013, adapted from Cohen, 1979a and Fjellman, 1992

continuum, or a hierarchy, but in gradations. Lovell concludes that there is a wider variety of forms of staging which are apparent in most locations, from acts of restoration and conservation in heritage sites to creative placemaking in cultural destinations, and this is a theme upon which this book will expand. Lovell suggests a rereading of Cohen's 'impressions', which could be further enriched by Fjellman's (1992) axiomatic categories used to describe Walt Disney World: 'real real, fake real, real fake, fake fake' (ibid., p. 255). The cubic format suggests dimensions of experience and the borders dividing the cubes are porous, indicating permeability between states (see Figure 1.2 for a more mobile method of modelling authenticity). Lovell went on to add 'hyperreal' to the authentitypes; this book also adds categories of 'unreal' and 'magically real'.

Fjellman's categories of fake and real are heuristic devices, but they help us further understand the relationship between authenticity and places. Lovell's research (see Tables 1.2 and 1.3) suggests that the 'real-real' experiences are hot, existential, personal and contemplative. 'Fake-real' states are cool, cognitive, exploring 'mechanistic authenticity', the facts beneath the surface of objects. 'Real-fake' impersonations are concerned with re-enactment, irony, performance and staging. 'Fake-fake' awareness of commodification can lead to feelings of alienation and hypervigilance. Hyperreality, magical reality and unreality are discussed later in this chapter; they all involve the sensation that reality is magnified, copies are real and reality copied (Eco, 1986; Baudrillard, 1981). Lovell's research indicates that all states may be affective and tourists switch between them, often rapidly, cool authenticity warming, or even superheating. One state can lead to another; for example, wonder about the impossibly accurate restoration of a building may lead to hyper-awareness of the extent of reconstructive staging in the historic environment. The state of mind, belief and knowledge of

Table 1.2 Real and fake experiences

States	*Sensations*
Real-real	Heating, existential, elevated, embodied, contemplative, inhabiting the past, sublime, performative, psychoheritage
Fake-real	Cooling, mechanistic, factual, distinction, status
Real-fake	Warming, enjoyment of performance, nostalgia, pretence, re-enactment, irony
Fake-fake	Freezing, critical, alienated, disappointed
Magically real	Evaporating, enchanted, transported, poetic, imaginative, the otherworldliness of the everyday
Unreal	Superheating, overwhelmed by reality which seems magnified, sublime
Hyperreal	Melting, 'wobble' when reality seems to be a copy

Table 1.3 Real and fake qualities

States	*Qualities*
Real-real	Entropic, original, ruined, gritty, slow
Fake-real	Historically accurate reproduction, first-order simulacra
Real-fake	Obvious copies, pastiche, postmodern references, re-enactments, recreations of historic or fictional places, second-order simulacra, fast, themed, decontexualised
Fake-fake	Poor copies of copies, commodified, placeless, post-truth
Magically real	Third-order simulacra, dreamlike, imaginative, rescaled, phantasmagorical, uncanny
Hyperreal	Overwhelmingly copied
Unreal	Singularity, iconic
Virtually real	Circulated, viral, holographic, immersive

tourists, or the 'faith in fakes' (Eco, 1986) when encountering places affects sensations induced by (and the placebo effects of) reproductions.

Cubes and boxes do not illustrate the permeability, stretching, movement, flow and elasticity of authenticity. Combinations of the terms fake and real are used throughout this book to explain the malleability of our experiences of copies and their qualities (as shown in Tables 1.2 and 1.3). Baudrillard (1981) discussed four stages of the sign order; the faithful copy, the unfaithful copy, the copy with no original and the simulacrum preceding reality. These differentiations can be applied to places, for examples, fake-real buildings, which are intended to be historically accurate copies or restorations, real-fake reproductions or pastiche and magically real art installations. Places may move category and/or occupy different categories simultaneously. Lastly, as the book discusses, most places, however copied or commodified, have their own truth.

Lovell's (2013) concept of the cycle of 'real to fake to real to virtually real' (Figure 1.3) further illustrates the sequential movement between categories. Some placemaking involves creating a copy, a reproduction, or a recreation, or a new way to frame places, which is resisted and then takes on different meanings and a life of its own.

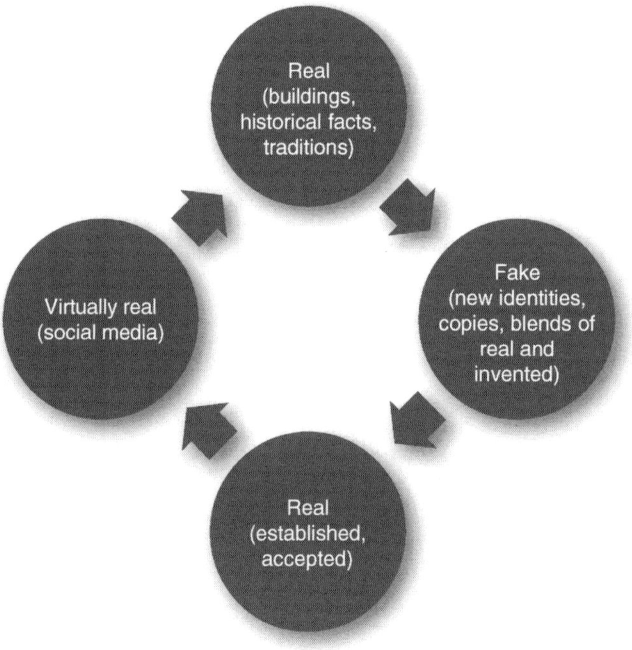

Figure 1.3 Real to fake to real to virtually real

Objective authenticity

One of the first pieces of research on the applicable and practical uses of objective authenticity was the measurement of the originality of museum objects by curators (Trilling, 1972, cited in Wang, 1999) and the idea of the curatorial pursuit of authenticity was supported by Smith (1989, p. 18): 'The notion of authenticity, the idea of the true cross, lies at the heart of all museum activity.' Bruner (1994, p. 401) also discussed museums and authenticity and observed four distinct categories of authenticity practised by museum practitioners in New Salem: 'verisimilitude, genuineness, originality and authority'. Ram *et al.* (2015, p. 111) state that the word 'authentic' is often described in terms of its characteristics being real, reliable, trustworthy, original, first hand, true in substance and prototypical as opposed to copied, reproduced or 'done the same way as an original'. Cohen (2007, p. 76) has highlighted how the concept has been deployed in tourism studies in several overlapping senses to include authenticity as 'origins', 'genuineness', 'pristinity', 'sincerity', 'creativity' and as 'flow of life'. Cohen implies that authentic sites are outside MacCannell's (1989) theory of the tourism sacralisation process of 'naming, framing and elevation, enshrinement and reproduction' of attractions. In relation to this last meaning, Cohen states that the quintessential authentic site is one that is utterly unmarked; it is authentic, precisely because it is not an attraction.

While the concept has been questioned by Reisinger and Steiner (2005), who have argued that objective authenticity is obsolete, Belhassen and Caton (2006) contend that it is both practised and consumed by a community of practitioners and tourists. In fact, as Lovell (2013) notes, acts of authentication on the part of tourists and conservators continuously co-create historic cities, which works reiteratively down to a granular level of detail. Drawing on Selwyn's (1996) categories, objective authenticity can be said to be cooler, relating to scientific and official forms of authenticity (Cohen and Cohen, 2012; Lovell, 2013). One of the most prominent ways in which objective authenticity can be seen to be operating is through official authentication of places – for example, World Heritage Site designation, the establishment of conservation areas and listed buildings and the designation of wilderness areas and national parks.

Constructive authenticity

Constructivists and social constructivists in the fields of tourism anthropology (Bruner, 1994; Cohen, 1988b; Smith, 1989; Edwards, 1996) suggest that there is no single reality, the mind attributes belief and assigns significance and therefore authenticity is negotiable, relative and subject to social trends. In this respect, 'tourists are seeking "signs of authenticity" or *symbolic* authenticity' (Wang, 2000, p. 49) at the same time as objects are constructed to appear authentic through 'images, stereotypes, expectations and power' (Wearing *et al.*, 2010, p. 29). Cohen (1988b) alludes to changing authenticity as 'emergent', as values and tastes alter, illustrated by his study of the commoditisation and consumption of African art. This view is shared by Bruner (1994, p. 407), who stated: 'all cultures continually invent and reinvent themselves', implying that places are in a constant state of flux and adaptation, never finished, they are always developing, an assemblage of connectedness, forces and flows (Massey, 2005; Harrison, 2013). Emergent authenticity is particularly evident in niche tourism – for example, the growth of dark tourism attractions (MacDonald, 2009) and industrial heritage during the twentieth century. Cohen (2007, p. 78) also cites Disneyland as an example of this. As he points out, what is generally regarded as 'the initially quintessential, overtly "framed" tourist attraction, has over time been increasingly recognised as part of contemporary American culture, and as such, as authentic'.

Hot authentication, existential and performative authenticity

One of the key ways to understand the authenticity of places is through experiences; hot authenticity is described by Selwyn (1996) as kinaesthetic, emotive and affective. Cohen and Cohen (2012, p.1300) built on Selwyn's typology to develop authentication as:

> an immanent, reiterative, informal performative process of creating, preserving and reinforcing an object's, site's or event's authenticity. It is typically an

> anonymous course of action, lacking a well-recognized authenticating agent. The process of 'hot' authentication is emotionally loaded, based on belief, rather than proof, and is therefore largely immune to external criticism.

Hot authenticity suggests that places can be consumed not just cognitively and intellectually, but also emotionally through 'personal authenticity' (Wang, 1999) – for example, private memories of an ancestral home or scene of a family event. There is a spontaneity to hot authenticity, a sense of Benjamin's (1936, cited in Harrison *et al.*, 2008, p. 5) 'involuntary memory', which can be a very powerful sensation, occurring when a sensory experience acts as a trigger to reanimate a memory – for example, Proust's Madeleine. Some places, such as heritage sites, also mesh with 'public memory' (Benton and Cecil, 2010, p. 1) and internalise collective memory to create their individual experiences, commemorating past events in dark sites such as battlefields. A further type of hot authenticity has been described by Relph (1976 [2008], pp. 51–4) as 'empathic insideness… a willingness to be open to significances of place, to feel it, to know and respect its symbols'. This category includes place attachment, incorporating strong feelings of identity which could arise from an experience of a global landmark; tourists are not simply grazing and gazing in an 'incidental' manner, there is the possibility of a deeper connection. Lynch (1972, p. 40) explores the emotional attachment of people to the built environment as personal, genuine, meaningful and symbolic: 'We have the sense that we and our works will also reach uninterrupted old age…. After a catastrophe, the restoration of the symbolic centre of community life is a matter of urgency.' This phenomenon was demonstrated by the 9/11 memorial and museum (Message, 2006, p.107), a new landmark acting as a collective memory site, visited with the intention of provoking emotional catharsis and understanding of the catastrophe.

It could be argued that the hot authentication of places can be superheated – for example, Tuan (1974) and Relph (1976 [2008], p. 37) cite 'topophilia', where the encounter with the place is profound. Lovell (2013, p.190) develops Csíkszentmihályi and Csíkszentmihályi's (1988) term 'flow' in sports psychology to a state of 'heritage flow' in heritage cities, which resembles peak sports experiences. In this elevated state of mind, the tourist can transcend the present to connect to 'deep history' or psychoheritage. Places and moments can 'strike a sublime chord as a reaction, not simply to physical grandeur, but also to a sense of the insignificance of self in the immensity of time'. Authentic flow can be argued to range from mindfulness to moods like tranquillity to feelings of being overwhelmed by the sublime when confronted by, for example, the age of a prehistoric monument. Places such as the wilderness are sublime in their grandeur, the sheer scope of the desert is overwhelming, a cathedral can seem spiritual and the vast enclosed space within offers a sublime authenticity. Time is diminished into a sense of timelessness and the building, mountain, or other colossal site or scene gives the viewer a sense of immortality and wonder.

Cohen and Cohen (2012) observe that hot authenticity can lead to existential authenticity (Jamal and Hill, 2002; Kim and Jamal, 2007; Steiner and Reisinger,

2006; Wang, 1999). In *Being and Nothingness*, Sartre (1943) attested that existence preceded essence – that people were in the world as entities before they were defined or defined themselves. He argued that to act authentically was to reject bad faith which restricted freedom of oneself, one's lifestyle or others. Existential tourism was pursued by the Beatniks in the 1950s, whose ideas of travel as hobos on railroads across America reflected their sense of authentic freedom, as described by Kerouac in 1957 in his novel *On the Road*. He compared his writing and their experiences to improvised, freeform jazz. Whereas objective and constructed authenticities are, basically, concerned with objects (and places), 'existential authenticity' is essentially activity related and can be described as a state of being, involving 'personal or intersubjective feelings activated by the liminal process of tourist activities' whereby 'they are engaging in non-ordinary activities, free from the constraints of the daily' (Wang, 1999, pp. 351–2).

As Reisinger and Steiner (2005) observe, this version of authenticity is bound up with a search for the 'authentic self', as Heidegger encapsulated in the concept of Dasein, a state of being-in-the-world. The authentic discovery of self for tourists can be linked to a search for adventure and related tourism activities associated with wilderness and the great outdoors. Furthermore, Wang's suggestion that existential authenticity may also involve a liminal process has relevance to the beach holiday, where a 'natural' liminal zone provides an environment for playfulness (Shaw and Williams, 2004, p. 139).

The 'performance turn' (Haldrup and Larsen, 2012; Urry and Larsen, 2011) has illuminated new ways of considering authenticity and place. Aristotle's concept of mimesis has been employed by anthropologists such as Taussig (1993) to indicate copying behaviours and the repetition of ritual: copied behaviours blending with real. As Edensor (2001) suggested, tourism can be performed; Cohen and Cohen (2012) define acts of authentication as performative. Lovell (2013) argued that performative authenticity is demonstrated by tourists at heritage sites as they consciously perform the habitual acts and rituals of the past, such as drinking beer in a pilgrim's inn, thus reinscribing the acts in the places where others first performed them. Coleman and Eade (2004, p. 55) and Coleman (2008) examine the rituals of pilgrimage, observing that: 'at Walsingham, pilgrims retrace the actions of past pilgrims' in a form of mimesis, 'remixing' where there is 'little distinction between the self and what's imitated, linking people and past in physical movement'. Knudsen and Waade (2010, pp. 1–3) characterised performative authenticity as 'transformative' and, by this token, places may be transformed by performative authentication – for example, by atmosphere generated at re-enactment events, or through the mimetic reproduction of historical architectural styles, both of which enable visitors to reinhabit a form of the past.

Many pilgrimage sites are perceived as sacred due to the presence of a relic; as Shackley, (2001, p. 38) attests, 'A relic, whether a fragment of wall tile or a remnant of the body of a saint, is literally priceless; it is a way of coming into direct contact with a higher reality which can usually only be approached in prayer, meditation, worship or sacrament.' Tourism has been referred to as a 'sacred journey' (Graburn, 1989) to see a place, or a pilgrimage to see the sights

(Goffman, see MacCannell, 1973 [1999]) and destinations are authenticated by the aura of secular relics such as the Liberty Bell in Philadelphia, which have witnessed seminal events. Shackley (2001, p. 29) refers to sacred places such as Lourdes, where visitors included souvenir shops within an encompassing bubble of pilgrimage. Authentic experiences can require the ability to perceive places selectively (Crang, 1996, p. 429; Rojek, 1997, p. 53) ignoring staged authenticity (Cohen, 1979a; Lovell, 2013) and consuming places hedonistically: 'hedonic consumption acts are based not on what consumers know to be real but rather on what they desire reality to be' (Hirschman and Holbrook, 1982, p. 94). Thus, if authentication is superheated by belief, it is possible to have an authentic experience of an artificial place. Lovell (2013, p. 280) details 'the positive heritage placebo effect', where tourists are unaware of false fronts on buildings and experience them as Tudor, rather than Victorian.

Embodied authenticity

A further way of examining perceptions of the authenticity of tourist places is through embodiment, a form of experiential authenticity, where people are physically present, being-in-the-world and responding to the resonance of places. The 'tourist gaze' (Urry, 2002) initially seemed to objectify, leading to the authentication and romanticisation of places. Urry and Larsen (2011) modified the gaze to encompass multi-sensual aspects, altering the privileging of the 'visual' gaze to a 'multi-modal' (multi-sensory) embodied, performative approach. The gaze incorporates memory-making, imagination, smell, taste, texture and kinetics, which can make experiences of places affective, and it is increasingly aesthetic, leading to olfactory and auditory authenticities, the smell and soundscapes which can be disrupted by tourism. Some arguments suggest that places are not inanimate objects existing for the projection of authentic attributes, but have intersubjective agency (Obrador Pons, 2003) – for example, we behave differently on the beach to the way we act on mountains. This is indicated to be more than environmental psychology; some places do seem more inherently authentic than others. Spiritual places such as Uluru have powerful effects on people, a view supported by Harrison and Rose (2010) when discussing the Aboriginal perception of the material world, or 'sentient places'. Harrison goes on to qualify this in a later work (2013, p. 32), by discussing Strathern's idea of 'distributed agency' in actor-network theory. Actor-network theory implies that places are made up of both the physical and symbolic (Harrison, 2013; van der Duim, 2007; Selby, 2004, p. 150) and that places can influence the way people behave. A more-than-human world suggests that when people project authenticity onto places, there is a response. As Barthes (1979, p. 8) has described, 'the Eiffel Tower looks at Paris', and Schwenger (2006, pp. 37–40) contends, building on the work of Lacan, that objects gaze back at us in a way which is 'diffuse', 'a sense that seeing happens, though without an identifiable source'. Harrison and Rose (2010, p. 266) also argue that there is less of a boundary between the world 'out there' and people, as demonstrated in Lovell's (2013) work on quantum heritage cities, which unfold and appear as tourists pass through unfamiliar environments and the cathedral gazes down on the traveller. If authenticity is, indeed, projected by tourists, it is not projected onto a blank canvas.

Running parallel with these ideas, non-representational theory could be said to be a hedonic, authentic form of experiencing place; less self-conscious or self-regulated than performativity and linked more to Lovell's (2013, p. 188) 'real-real' experience comprising 'stream of consciousness, quantum space' and superheated and deepest authenticity. This is the practice of space which is creative, fluid and poetic (Bachelard, 1994; Massey, 2005; Relph, 1976 [2008]; Tuan, 1977). Thrift (2004) uses dance and performing arts theory and Obrador Pons (2003) refers to the act of lying on the beach to convey the body in space. Seamon (1980, p. 149) discusses 'natural attitude', the preconscious, phenomenological habitual movements, which he calls the 'space ballet'. Non-representational theory is applicable for the discussion of authentic experience because the lack of consciousness implied in non-representational movement connects people more intuitively to place. While the non-representational act of dance seems abstract, Edensor (2010) explores the act of walking as containing rhythms of space, creating a 'real' heritage experience and 'bodily intentionality' (Seamon, 1980), which means that not every movement needs to be considered consciously; tourists may be lost in their thoughts as they negotiate places and respond to them. Places are not static, or two-dimensional; they are known through a phenomenological 'place, practice, process', leading to a 'place ballet' (ibid.).

Postmodernism

A state of 'hyperreality' characterises the postmodern era, according to Baudrillard (1981) and Eco (1986), who suggest the indistinguishability of simulation and reality. Wang (1999, p. 356) suggests that postmodernists do not consider inauthenticity and its deconstruction a problem. Baudrillard (1981, p. 6) traced the four stages of the image, which were touched on earlier in the chapter, moving from reflecting reality, masking reality and masking the absence of reality, to bearing no relation to reality. He outlined three orders of the simulacra corresponding with eras of production. They consist of the first order, the copy (dating from premodern times); the second order, 'corrupting reality' (relating to modernism and mass-production techniques), and the third order (postmodern, which *precedes* reality), for example Walt Disney World (WDW). Eco's (1986) work on hyperreality also examines Disney theme parks, which are based on a mixture of fantasy and imagination and, as Wang (1999, p. 357) points out, 'in WDW there is no absolute boundary between real and fake'. The real may turn into the fake and vice versa and the 'lines between the real and the fake are systematically blurred' (Fjellman, 1992, p. 255). Wang (1999) also refers to work by Cohen (1995), who argues, as did Boorstin, that postmodern tourists have become less concerned with the original, pursuing instead a 'playful search for enjoyment' or an 'aesthetic enjoyment of surfaces'. Huxtable (1997, p. 81) maintains that recreated rainforests in Las Vegas are 'better than the real thing with added white tigers', which implies hyperreality is 'more real'. While the postmodernist approach might argue that many tourists are no longer concerned with the authentic, it is not without its critics; Lovell's (2013, p. 224) study suggested that in the context of historic cities, visitor behaviour is not one of post-authentication or postmodern authenticity; while authenticity is negotiable, tourists continue to search for relativity and authority.

Unreality

Unreality is related to hyperreality and is apparent in the reactions of some tourists to Canterbury Cathedral (ibid., p. 219), which seems unreal, overly real or, as one tourist commented, 'really real?' – too hyperreal to be true. Rojek (1997, p. 56) described Freud's 'disbelief' when he saw the Parthenon, which Freud ascribed to having seen so many pictures of the building that the three-dimensional structure was less tangible. In the same sense, people travelling to a destination which has been depicted on film and television, to the extent that it has become overly familiar, such as New York, describe a simulacrum.

Lovell (2013, p. 290) observes that rather than experiencing what Clarke (2005) describes as 'a cinematic city' – for example, the sensation of walking onto a film set when visiting New York – visitors to Canterbury felt as if they were entering a fairytale environment, applying the 'magical realist gaze' or 'Grimm gaze'. Gray (2006, p. 243) states: 'Think of Florence, Paris, London,

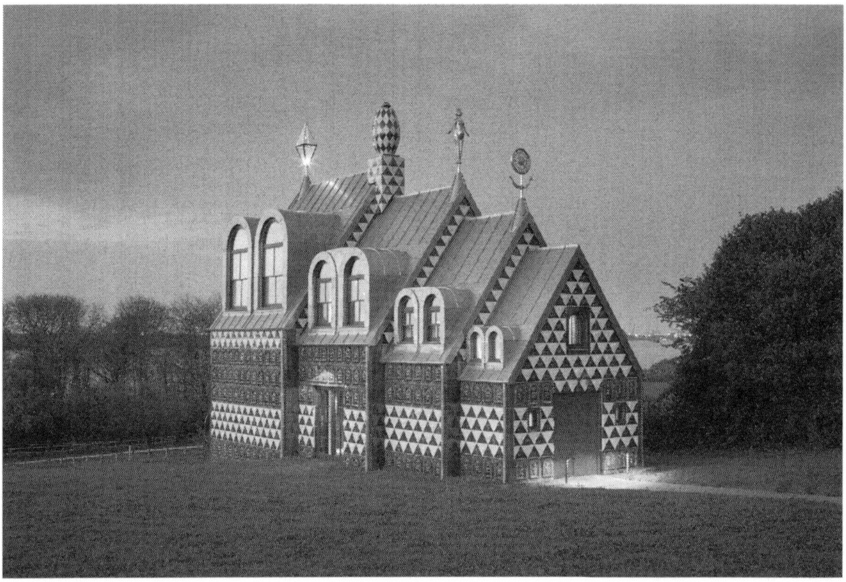

Figure 1.4 A House for Essex by Grayson Perry
Source: Jack Hobhouse

New York. Nobody visiting them for the first time is a stranger because he's already visited them in paintings, novels, history books and films.' Architecture throughout the ages has been inspired by romance and history, as demonstrated in gothic revival, or the castles of Ludwig II of Bavaria, who designed the romantic, fairytale Neuschwanstein Castle based on the refracted notions of the Middle Ages and the musical mythology of Lohengrin, as realised by Richard Wagner. Mediatisation and postmodernism can make historic structures such as castles and cathedrals seem more like fictional creations. Magical realism, as described by Faris (2004), can reveal the fantasy and poetic elements in the details of visited places. Examples of magic realism architecture encompass simulacra such as Gaudí's Sagrada Familia cathedral in Barcelona, which resembles a building from the pages of Dr Seuss, or Grayson Perry's House for Essex, which seems to mix elements of fairytale with stave church, or the iconic architectural works of Zaha Hadid.

Modelling authenticity

The study of authenticity involves navigating negotiability and encountering portmanteau words such as 'Tudorbethan' and contradictory ideas, combinatory forms, oxymorons such as 'genuine fakery' (Brown, 1996) and metaphors which are as thoroughly mixed as the multiple realities. The delight of researching authenticity is that it seems to stretch, turn, fold in on itself, unfold, emerge and retreat; these are the flows of authenticity within this text. It seems to the authors on writing this book that perceptions of authenticity form bubbles of liquid reality. Benjamin (1936 [2008], p. 7) describes how authenticity 'wobbles' and bubbles quiver, spin, replicate, stretch, join, divide, pop and create shapes when pushed together; they are permeable to some substances, forming structures under the surface and always moving in, of and around places. The model in Figure 1.5 is designed to suggest some of the multiple realities of places, which shape tourists' impressions.

The structure of the book

As indicated at the outset, this book aims to examine the authenticity of a range of places or environments that are important for tourism. The book is loosely divided into three sections: heritage, from the conservation of cities to gentrification; places which have been created to a greater or lesser extent by natural environmental processes, from seaside resorts to wilderness; and performing places, from event spaces and film locations to theme parks. Each chapter examines a place or environment and considers the variety of ways in which authenticity is perceived, refracted, or developed; the chapters are linked through theories of placemaking, including staging, theming and reproduction.

Chapter 2 begins by investigating how objective authenticity is practised in the development of English historic cities. It traces their journey as heritage

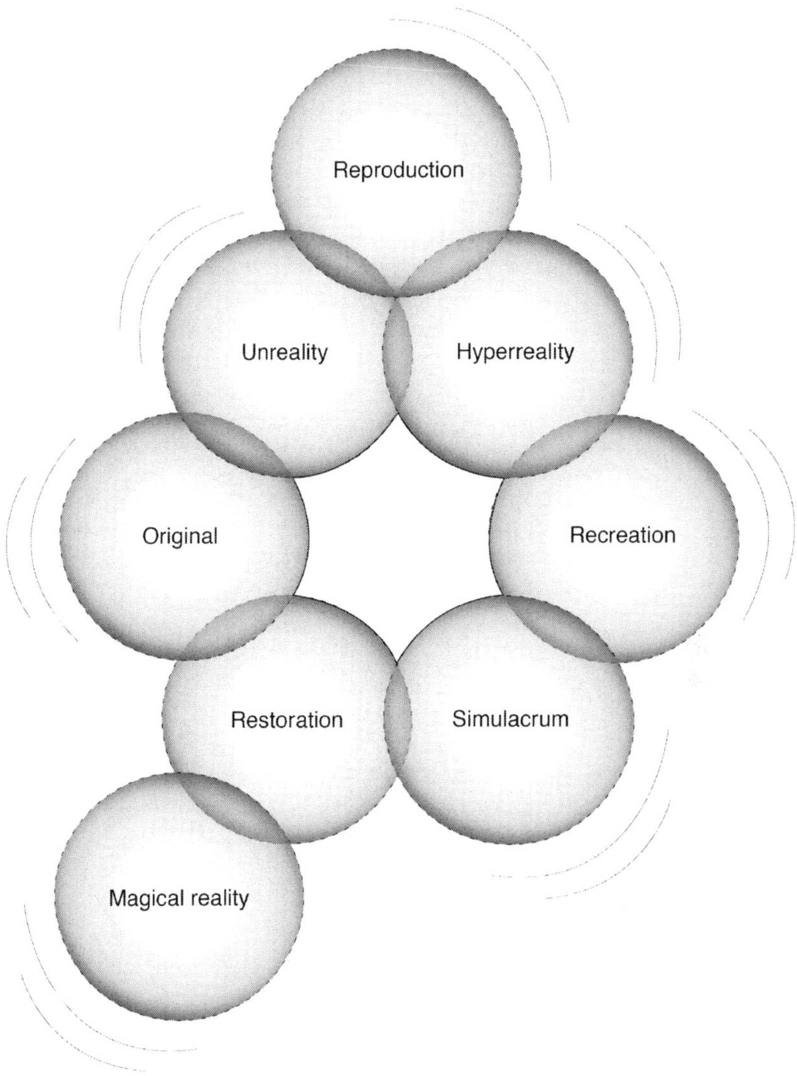

Figure 1.5 Model of realities

strongholds which came under siege from modernity after the Second World War, comparing their management to museum curatorship, focusing on the myriad subjective decisions involved in urban heritage planning – suggesting that objective authentication is a process and practice. The chapter then examines how tourists consume heritage cities as primary forms of interpretation, seeking the city of origin through various methods, such as finding a 'bird's eye view of time' from high buildings and exploring the streetscape. Lastly, the chapter uses ruins to illuminate the collapse and reconstruction of time within a 'real-real' experience.

Chapter 3 investigates approaches to restoration, including the theory of the Ship of Theseus, asking whether a completely restored artefact still retains the qualities of an original. Different forms of restoration and recreation are identified, from the large-scale, accurate recreation of some areas of cities, such as Gdansk, to de-restoration techniques, each of which has implications for the authenticity of places. Lovell's (2013) theory of 'mechanistic authenticity' involves the search for official signs of authentication, for example, by locating the age of artefacts. The chapter illustrates how cool authentication can be superheated by wonder at the skills of what seems to be an unbroken line of craftsmen stretching from the past to the present day. The chapter also assesses how explanations of the many architectural styles and false facades of buildings are experienced by tourists on guided tours, leading them to suspend their sense of disbelief and 'inauthenticate' places, searching for fakery.

The book next examines staging in the form of the nostalgic design tropes of gentrification. Gentrification could be said to be one of the ultimate forms of reproduction, with cities and neighbourhoods evolving from Zukin's (2010a) 'gritty cities' through what Lovell (2013) describes as a recalibration from 'real to fake to real', as working-class areas and bohemias are extinguished and 'neo-bohemias' (Lloyd, 2010) evolve. Chapter 4 contributes original ideas about the hipsterisation of places, centred on the adaptation, repurposing and commodification of industrial heritage and suggests that the mechanistic authentication of places occurs through the consumption of artisan products. Festival market places amplify the concept of heritage shopping, confusing the timescales by giving a heritage feel to modern consumption spaces, potentially clouding tourists' impressions of the age of places. The contemporary adoption of pseudo-history, theming and pastiche styles reflects postmodern design in the public realm, and can lead to a sensation of perpetually replicating environments. Gentrification is designed to appeal to a mainstream culture as it ripples outward and tourists seek new backstages.

The merging of magical reality, imagination and place in the staging of outdoor, site-specific art installations is explored in Chapter 5. The chapter suggests that tourist-flâneur encounters with hyperreal simulacra in the form of fantastical, surreal and outsize art may evoke dreamscapes, phantasmagorias and imaginaries. A variety of examples of creative placemaking illustrate the use of art as instrumentalism to underpin wellbeing and social capital and the chapter pinpoints how some landmark galleries and highly participative art aspire to locate creativity more authentically within communities.

The next section of the book examines places shaped by natural resources, first addressing the British seaside resort. Seaside resorts, discussed in Chapter 6, were established specifically to provide a place for pleasure and as an escape from, and contrast to, the drab and artificial (inauthentic) environments of urban centres in which most people lived and worked, suggesting that seaside visitors are searching for the authentic self. Resorts could be said to be cultural constructions, imbued with nostalgia and the chapter will examine these views as well as considering their liminal nature, especially in relation to the beach, and the links with 'existential authenticity'. Furthermore, resorts are dynamic places which, while containing relicts of the past, have also gone through a series of transformations over time making any assessment of their authenticity complex and blur any fakery. Chapter 7 on the 'rural idyll' interprets the English countryside as a culturally constructed place, suggesting that the simpler life is what we are all, essentially, pursuing, but that it doesn't exist. It may be suggested that the notion of the idyll and wilderness are the object of the constant search for authenticity which were discussed earlier. The chapter portrays a physical and social environment that is extremely attractive and superior in nearly every respect to that of the town and city and yet much of this image is at odds with reality. As such it is a classic example of Wang's (1999) 'constructed authenticity'. This chapter examines the nature and basis of the myth and reality, and assesses the power and influence of the idyll on people's lives and its impact on rural planning. The coverage then examines the nature of wilderness areas (Chapter 8), places which, at first glance, might suggest an obvious example of 'objective authenticity'. However, although some areas might be regarded as pristine, natural environments, and specific areas have been 'officially' designated as such, a deeper examination reveals that 'wilderness' is a complex and contestable concept which may also be culturally constructed. Furthermore, wilderness areas provide opportunities for a range of outdoor pursuits and life-enhancing experiences that equate with Wang's (ibid.) concept of 'existential authenticity'.

The next topic of the book concerns performed places. Places are singular, by their very nature, but sometimes they compromise that by trying to be somewhere else. Mass reproduction is a feature of Chapter 9 on film locations and photographed places, which investigates the idea of aura, mixed reality and the filmic gaze. The chapter squares the 'circle of representation' (Jenkins, 2003), examining how many places are changed to resemble films or photographs, replicating the enhanced settings, making fantasy reality. The chapter looks at the mechanistic authentication rituals of film tourists and draws a parallel with heritage re-enactors. The act of photography is synonymous with the tourist and reflects different aspects of authenticity such as the staged or the paparazzi shot. It is suggested that photographic images may be unreliable narratives and, in contrast, that there is a depth of poignant authenticity to be found in the 'punctum' (Barthes, 1980) or detail within the image. Chapter 10, on events, examines how events may sometimes function as a mask for places, giving them a transcient, dual identity. Creative placemaking may involve using events to transform spaces into stages, or art to add surreality, and these constructed elements lead to myriad experiences. Events are transient, but have the constructed power to totally transform places; hallmark events are synonymous with the places where they are held, community events may be deeply rooted in peripheral places, carnivals have the potential to create heterotopias and liminal spaces, involving communitas and ritual. Re-enactments involve a high level of authentication and some festivals begin as invented traditions, then, through emergent authenticity, move from fake to real. In addition to considering such issues, the chapter also examines hyperreal events and hyperfestivity, asking if events are too formulaic and ubiquitous. Finally, Chapter 11 on themed places explores a variety of environments that have been specifically constructed to create an artificial (or falsified) environment or image. Such places include Disneyland and the archetypal 'real-fake' city of Las Vegas, both of which betray hyperreality evident in their approach to commodification, as do the colossal shopping malls also addressed in this section. The chapter assesses how museums have been compared to theme parks, an increasingly relevant paradigm in their use of immersion and atmospherics to engage visitors and merchandise which reproduces artworks and artefacts and the digitisations of their collections. It is not that everywhere is authentic; the work suggests that sometimes there are two levels of belief operating; for example, a piece of furniture in a theme park is designed to look very old, the mind understands that it is not old, but the viewer chooses to behave as if it is a museum piece. While themed places seem to be artificial, the chapter debates how tourists authenticate the inauthentic in the most fantastical of places. On occasion, we enjoy it when places assume a cloak of authenticity or a mask of inauthenticity. The book suggests that, far from being post-authentic, tourists take pleasure in distinguishing between shades of real, magically real, hyperreal or fake, all of which matter, now more than ever in the age of post-truth.

2 Historic cities

From museumification to the origin of the spaces

Introduction

Discussions of authenticity frequently focus on the extent to which objects and places are original. The chapter draws from some of the findings of Lovell's (2013) research study of staged authenticity in 15 English historic towns and cities: Canterbury, Oxford, Cambridge, Chester, Durham, Bath, Lancaster, Chichester, Shrewsbury, Winchester, Salisbury, Brighton, Rochester, Lincoln and York. The European Association of Historic Towns and Regions (2002) defines historic towns and cities as environments 'including villages, small towns, cities and parts of larger urban areas with significant cultural and heritage assets'. The association adds: 'All definitions, however, should include as a minimum requirement that each region, town, city or village has an historic centre which has evolved over the centuries and whose character is still intact.' The examples used in this chapter are applicable to a global range of historic environments, many of which share common issues and tensions between development and conservation. Lovell (2013) explored how the aura of historic cities is linked to what Heidegger (1971, p. 165) describes as the 'gift', which is to communicate a heightened sense of the 'flow of time' (Lynch, 1972, p. 44). Many historic cities have been uniquely preserved as urban environments, by a resistant, often rebellious, conservation ethos. Although some historic cities are walled, they are not bounded, unlike many relatively contained and controlled heritage visitor attractions. They epitomise organic, changing, living history, subject to flows of ideas and capital and open to a variety of visitors, creative influences and private sector industries, all of which help to construct their identity.

This chapter investigates how a range of actors who co-create places apply 'objective authenticity'(Wang, 1999, p.351). Harrison (2013) explores the complexity of heritage actor-networks and Harrison and Rose (2010, pp. 266–9, emphasis in original) argue that: 'there is no such thing as intangible heritage without tangible objects and, more importantly, *places* to attach itself to'. Wang (1999, p. 351) described the estimation of authenticity by museum curators as the 'absolute and objective criterion'. The curatorial pursuit of authenticity was supported by Smith (1989, p. 18): 'The notion of authenticity, the idea of the true cross, lies at the heart of all museum activity.'

Bruner (1994, p. 401) has also added to the debate, observing four distinct categories of authenticity referenced by museum practitioners in New Salem: 'verisimilitude, genuineness, originality and authority'. Reisinger and Steiner (2005) argued that objective authenticity is obsolete, but, as Belhassen and Caton (2006) contend and this chapter also investigates, objective authentication is both practised and consumed by a community of placemakers and tourists. As Relph (1976 [2008], p. 71) details, historic sites are socially engineered and the monumental walls, street layouts and key buildings both suggest and conceal the people behind them.

This chapter explores how people authenticate places, locating the 'original cities' through the layers of modernity, excavating beneath the modern carapace. Lovell (2013, p. 78) examined a range of tourist experiences of staged authenticity, including (drawing on the categories used by Fjellman [1992] to explore Disney World) 'real-real' impressions of historic cities. Lovell suggests that real-real experiences are spontaneous and, in some cases, encompass a state of 'heritage flow', where the aesthetics of the historic environment are appropriated without analysis – for example, when encountering a ruin. Heritage sites and cities can provide three-dimensional, evidential topographies of memory (Jordan, 2006; Harrison *et al.*, 2008) and the chapter examines how acts of authentication in the network of material heritage, performed by tourist and conservator actors, affect the aura of historic cities, by creating embodied collective memories. Harrison and Rose (2010, p. 266) contend that people and things 'bind time, place and generations… keeping the past alive in the present'. Preserving heritage environments imbues them with meaning and layers of authentication which supersede their economic value.

A history of historic cities

Paradoxically, a key stage of the more recent 'journey' of some English historic cities to their contemporary state was the Second World War Baedeker bombing raid (West and Ansell, 2010) which responded to the Royal Air Force bombing raids of Germany that had begun with the destruction of Lübeck. An urban planner described the impact of the bombing thus:

> The pre-war period was benign, a time of romantic conservation; Canterbury development bylaws allowed buildings through, but there wasn't widespread destruction. The war was the fault-line and a source of trauma; up to 25% of Canterbury was destroyed during the war.
>
> Lovell, 2013, p. 17

While the Victorians had destroyed and appropriated a great deal of building stock with their wealth and desire to modernise, the Baedeker raids were pivotal.

After the Second World War, 'social and economic changes accelerated the physical destruction in the historic environment' (Smith, 2006, p. 100). The new

Elizabethan age and Festival of Britain were part of a modernisation programme, which included a Georgian revival period, followed by the influence of Scandinavian flat-roof design and American modernism in the 1960s, with the widespread use of concrete. Developments with no historic references or anti-historic geometric design styles were erected in new materials such as concrete in bombed areas of historic cities – for example Canterbury's brutalist multistorey car park. The wear on the concrete meant that many buildings were quickly no longer fit for purpose. Although these buildings often had functional value, they strongly contrasted with the heritage aesthetic and the mimesis of restoration, which perpetuated nostalgic architectural styles. In addition, after the war, various road schemes were proposed which would involve the large-scale destruction of key ancient areas in historic cities. The proposals meant that many structures were neglected in anticipation of demolition; therefore, some local authorities built up large portfolios of buildings through compulsory purchase.

As modernisation continued in the post-war era, many British historic cities began a rolling conservation programme, aiming to preserve the past as an antidote to modernism. Historic buildings began to be listed in 1949, giving buildings 'new, universal layers of meaning' (ibid., p. 101) and the 1971 Town and Country Planning Act meant listing took place on a wide scale. UNESCO had produced the key International Charter for the Conservation and Restoration of Monuments and Sites, the Venice Charter, in 1964. As Samuel (1994) described, Conservation Areas were also established in the 1960s and 'buildings at risk' lists were compiled in Chester and York in the early 1970s. The Ministry of Public Buildings and Works was a product of the Ministry of National Heritage Act of 1983 which established English Heritage to manage the national built heritage, including the distribution of grant schemes. Preservation ethics took hold in the 1960s 'as a pole of opposition to the modernisations of the time' (Samuel, 1994, p. 129), with the reappraisal of vernacular architecture, and the folk movement. Authenticity continued to be popular; during the 1970s 'rescue' archaeology developed, as did the nostalgia-fuelled trend of reappraised authenticity or 'Retrochic' (ibid., p. 83).

Historic cities display many attributes of constructed authenticity. The 'concept city' or Second Space (de Certeau, 1984; Donald, 1992; Lefebvre, 1991; Soja, 1999) is the preserve of planners and architects, who try to make sense of place by gaining a vantage point, an insider view, looking down on the whole. Planning in historic cities has been defined as consisting of four stages in Ashworth and Tunbridge's (2000) description of the 'tourist-historic city'. The first stage is characterised by the preservation of heritage buildings, with extra policy control. Second, there are functional changes, including the migration of businesses into and out of the heritage area. Third, there is a separation of the historic from the commercial city and, fourth, a partial physical separation and overlap of the commercial and heritage areas (Ashworth *et al.*, 2000, p. 210). The original tourist-historic city model doesn't truly address the contemporary reappraisal of the value of 'heritage cultural capital' (Sorkin, 1992) where heritage, culture, creativity and enterprise are synonymous and what Richards and

Wilson (2007, p. 13) describe as 'heritage mining' combines intangible cultural capital with the fixed capital of the city. The 'tourist-historic city' approach is limited because it only deals with geographical space; in contrast, Relph (1976 [2008]) identifies several different spaces, including geographical, existential, sacred, cognitive and abstract, and approaches them holistically to achieve place identity. The 'tourist-historic city' conceived in the 1990s (although revised in 2000) was materialist, reflecting the hypotheses of the time, and doesn't fully take account of these other spaces (or places). It reflects a time when historic cities were 'honeypots', to the extent that Canterbury was receiving up to 18,000 coaches per year and actively participated in de-marketing practices. At that time, the primary research on historic cities focused on the fragility of the environment, and visitor management issues (Grant *et al.*, 1999a, 1999b, 2000) and crystallised in Russo's (2002) 'vicious circle' of tourism development. Since that time, cultural imperatives have gained greater currency (Stuart-Hoyle and Lovell, 2007) and the discussion of heritage moved on to subjects such as gentrification, dissonance and ownership (Zukin, 2010a).

Conservation bodies involved in historic cities can be diverse and multi-platform. Investigating Berlin, Jordan (2006, pp. 83–9) addresses the deeper complexity of urban heritage memorialisation, where, rather than a collective identity, heritage sites and markers reveal the diversity of those involved with 'public funds concentrated in certain districts' and market forces all playing a strong part in the process of forgetting or preserving: 'These sites are adapted, altered, reused, modified, updated, and renovated by a diverse array of groups and individuals, working in conjunction with officials (elected and otherwise)' (ibid.). Planning decisions in historic cities, and therefore nominal control, rests with local authorities and English Heritage, but this is only a part of the picture. Heritage placemakers include a vast network of private and public organisations with a vast array of different objectives, from conservation to economic development, structural engineering, cultural placemaking and entrepreneurship, to name just a few.

Amenity groups such as the Canterbury Society had a strong voice in the development of historic cities; they formed the Citizens Defence League, which took seats on the Council in the local elections to stop developments such as the Canterbury road scheme, which was abandoned in 1977. In some senses, it could be contended that an enduring 'rescue' ethos pervaded historic city culture, as epitomised by Canterbury architect Anthony Swaine. He practised from the post-war period to 2012 and allegedly chained himself to railings in Canterbury Cathedral to prevent iron from the tomb of Thomas Becket being taken away to be melted and used for tanks. Ideologies of rescue amount to a resistance to modernity and the anthropomorphisation of the plight of historic cities established a modus operandi which has lasted for decades. This has been demonstrated by resistance to new architectural projects in Bath, Brighton and Plymouth, and was epitomised by the 2010 Save Our Museums campaign in Canterbury, formed when the local authority threatened the Westgate Towers Museum with closure. The Nara Document on Authenticity (UNESCO, 1994) underlines the concept of preservation as a moral imperative:

> In a world that is increasingly subject to the forces of globalization and homogenization, and in a world in which the search for cultural identity is sometimes pursued through aggressive nationalism and the suppression of the cultures of minorities, the essential contribution made by the consideration of authenticity in conservation practice is to clarify and illuminate the collective memory of humanity.

Museology and the conservation of the built environment have much in common; it could be said that they are not passive, but active, performative, cultural acts of memory-making. The concept of the 'museumification' (Wright, 1985, cited in Harrison, 2010) of heritage cities is not used here in the pejorative sense, but as collections of heritage objects; historic cities resemble museums of artefacts. As early as 1775, parallels were drawn between cities and museums by Boswell: 'I see a vast museum of all objects, and I think with a kind of wonder that I see it for nothing' (Boswell, 1775, cited in Kirshenblatt-Gimblett, 1998, p. 51). There is evidence of the objectification of historic cities in a discussion of the tourist-historic city, where Ashworth and Tunbridge (2000, p. 135) describe the 'historic gem' city and Boyer (1996, p. 34) refers to 'the city as a work of art', a design approach persisting until the end of the nineteenth century, with collections of buildings termed 'jewels'. Preservation creates cities which are put on display and performed, making conservation a form of staging as gentrification and cultural placemaking gain a stronger hold.

Conservation: memory and time

One of the most prominent ways in which 'objective authenticity' can be applied at city-wide level is through official authentication – for example, World Heritage Site designation in cities such as Canterbury and Bath. UNESCO and ICOMOS are largely responsible for establishing global heritage practice (Askew, 2010, p. 26) and heritage practitioners influence and respond to documents such as the Nara Document on Authenticity (UNESCO, 1994): 'The understanding of authenticity plays a fundamental role in all scientific studies of the cultural heritage' (ibid.). A Statement of Authenticity/Integrity is produced as part of the World Heritage Site application process (UNESCO Operational Guidelines for the Implementation of the World Heritage Convention, 2008), taking the form of a generalised declaration of the state of site authenticity in terms of threats from external development pressures. The emphasis on 'objective' or official authentication has practical implications when applied on a city-wide basis; designations such as Conservation Area status make it more difficult to modernise, remove or redesign listed buildings. Critical mass – for example, 2,700+ buildings in Canterbury, or World Heritage City designation in Bath or Edinburgh – creates a sense of 'preserved place ethos', in addition to design considerations. Objective authenticity could be used to validate the provenance and list buildings in historic cities, using empirical methods.

Conservators and planners appropriate and authenticate historic cities, viewing them with a multi-modal 'heritage gaze' (Urry and Larsen, 2011; Smith, 2006). They strive to perpetuate certain qualities of objective authenticity, including local or traditional production, skills and materials, all of which enhance the aura of places. When considering aesthetic judgement and taste, Otero-Pailos *et al.* (2010, p. 61) supports Fitch's (1990) argument that restoration 'is more of an art than a science'. For example, there has been a shift in taste in historic cities in the late twentieth century, from preserving the more 'polite' civic buildings, towards what Fairclough (2008, p. 298) describes as more 'vernacular' architecture. The heritage tourist would experience not simply the current patina of age in historic cities, but the traces of many past decisions about preservation and conservation. The process of conservation is (unconsciously) selective, requiring decisions about what is worth preserving to remember, or what to ignore, or forget. Memory itself is subjective (Smith, 2006, p. 58) and unreliable; collective memory is also constructed. Tunbridge and Ashworth argue that historic cities can be designed to accentuate aspects, perhaps eras, of the past: in the practice of 'heritage planning' (Tunbridge and Ashworth, 1996, p. 24) or 'purposeful preservation' (Ashworth and Tunbridge, 2000, p. 9), which has been 'objectively' guided by conservation criteria. Lynch (1972) makes the point that, perhaps, in practical terms, the cheaper the material, the less likely it is to age aesthetically, therefore the expensive stone structures of the affluent survive and lead to a built heritage of the wealthy: 'Paradoxically, those places in which local "concern for the past" is often so marked among successive generations of moneyed and leisured inhabitants, actually tell one less about the past as a whole… than do more ordinary, battered places' (Tindall, cited in Lowenthal, 1985, p. 243). It could be argued that historic environments are associated with intellectual and national cachet and status (Bourdieu, 1984). Authentication practices at global, national and local level foster collective memory, reflecting 'particular ideologies' (Salazar, 2010, p. 130; Maitland and Smith, 2009, p. 172) which can be aesthetic, or political – for example, in the case of Nuremberg (MacDonald, 2009). There is as much forgetting as there is remembering in the contemporary heritage landscape of historic cities.

It has been suggested that historic cities have been socially and physically held back by conservation. Coates (2000, p. 222) describes the way historic cities have been '"heritagised"… suppressing the chance danger and unpredictable experience that makes the essence of the urban experience'. Smith (2003, p. 24) portrays historic environments as time capsules, and Boyer (1992, p. 1) as tableaux. External forces which could have precipitated change in other types of urban environment such as post-industrial cities or seaside resorts are less evident in fortified or walled cities, which are often still partially bounded or marked by gates and widespread ancient buildings. Historic cities were built for smaller populations; the ratio of population to tourists is low, the historic core tends to be concentrated and compact, making it more 'conservable', and the exercise of objective authenticity has been shaped and slowed by regulatory devices such as Conservation Area and Listed Building status (Breakell and Human, 1999; Grant *et al.*, 1999a, 1999b, 2000).

While English Heritage controls the lists and has been accused of setting England's heritage 'in aspic', Simon Thurley, Head of English Heritage, pointed out, refuting this idea, that aspic is not even used any more (Thurley, 2010). There is an underlying assumption about a consensus in historic cities in favour of heritage preservation. As described in Lovell (2013, p. 264), this comment by a historic city placemaker demonstrates how they have evolved from a position as conservation strongholds:

> there was a very strong view that the city needed to be protected against the unpleasant and unwelcome sort of modern approach... he used the phrase to me 'we're fighting a war'... What I want is when the opportunity comes up is to add the 21st century in a good way that compliments the city as opposed to replacing the city.

When consuming historic cities, the tourist may search for architectural narratives. Historic cities can be associated with a predominant sense of place genre, such as Georgian Bath, but they are rarely of a single style. Lynch (1972, p. 43) described the idea of the urban environment as a 'public attic' jumble of eras, an evolving collection of styles, evocative, old-fashioned and anachronistic, reflecting the tastes and mores of past societies. Looking beyond geographical space to the longitudinal 'heritage space', from Bronze Age settlements to contemporary eras, historic cities are an organic, dynamic, changing entity. Lynch (ibid., pp. 135–62) lists the references to changing time while walking in the historic city of Boston: the clocks, the signs such as parking meters, with times on them, the contrast of old and new, even the trees which change with the seasons, indicating the time of year. One of the effects of this assemblage, or 'time collage' (Schlereth, cited in Lippard, 1997, p. 93) is to lend the city a visible history, the cumulative effect of which reinforces the sense of authenticity; as Smith (1989, p. 20) states 'the life-cycle of an artefact is its most important property'.

Ancient buildings can be thought of as relics, used to stimulate collective memory (Chronis, 2006, p. 269), or 'artefactual memory' (Urry, 1995a, p. 219), providing sacred sites for what could be said to amount to a 'civil religion' (Bellah, cited in Loewen, 1999, pp. 25–6). The movement towards preserving and generating heritage combines with shifts towards speculation, spectacle and a growing 'historical culture' (Bagnall, 1996, p. 229). Hewison (1987) mounted a pejorative attack on the 'heritage industry', as did Huxtable (1997, p. 22): 'What are we to do about the past? Continue to exploit and distort it, turning it into a caricature and a crowd-pleaser while pointing piously to what we have "saved"?' In a similar vein, Lynch (1972, p. 35) also questioned the validity of the conservation movement and the process of preserving the past. This was echoed by other authors such as Lowenthal (1998, p. 94), who stated, 'everything the government calls heritage is holy' or 'no matter what comes up in the town, someone fastens the word *heritage* to it'. The word heritage could now be substituted for the all-pervasive use of the adjective *authentic*.

Cities of origin

Lovell (2013) argued that real-real experiences of places are instinctive and include the way in which tourists often attempt to locate the original city beneath the contemporary surface by ascending to the highest point in the landscape, gaining a 'bird's eye' point of view. They orientate themselves by exploring the founding, shape and growth of places. De Certeau (1984) described the view from the World Trade Centre as highly utopic and conceptual. Expounding on the view from the Eiffel Tower, Barthes (1979, p. 11) states that this:

> generally intellectual character of the panoramic vision is further attested by the following phenomenon, which Hugo and Michelet had moreover made into the mainspring of their bird's eye views: to perceive Paris from above as infallible is to imagine a history.

He continues (ibid., p. 4):

> the Tower acquires a new power: an object when we look at it, it becomes a lookout in its turn when we visit it. Glance, object, symbol, such is the infinite circuit of functions which permits it always to be something other and something much more than the Eiffel Tower.

Lovell (2013) detailed the view from the hills of Durham or Lancaster, the panorama from the castle at Rochester and, notably, the view of Bath cathedral from the roof pool of the new Bath Spa, which reflects how people try to encompass and convey the conceptual city in photography. The panoramas captured in Lovell's study feature a number of rooftops and silhouetted monumental urban skylines aimed to reinforce distance and aura.

The bird's eye view is not simply visual; it links to the imagination, spanning time as well as space, indicating a state of what Lovell (ibid., p. 190) terms 'heritage flow', encompassing shared collective memory, indicated by one placemaker when discussing 'our Roman past'. One of the distinctions of historic cities is their emphasis on distant 'far-off' time periods and ancient history, which compounds their aura and sense of mystery. There is an awareness in contemporary planning of a strong sense of responsibility for maintaining the longitudinal continuity of place authenticity. In historic districts the planning emphasis is multilayered, simultaneously linking highly diverse timescales. At the Royal Observatory in Greenwich, a placemaker described changes to be made for the coming Olympic Games. As few of the developments had been started at that time, this was a conceptual 'bird's eye view of time demonstrating a vista encompassing an imagined Greenwich or future dreaming'. Here, memory-making was rooted in what Crang (1997) described as 'future memory' and the view was multilayered, observed from the observatory, or 'place where time began'.

Original places are both visible through the bird's eye view and detectable at pedestrian level, not only in the form of the geography, hills and waterways, but also in the contours of the architecture and streetscape, which, over time, have become part of the natural topography. Lovell (2013, p. 191) elucidates how waterways – which served as arteries, feeding the first settlements of many historic cities – are observed and recorded as such by many tourists and placemakers in Bath, Rochester, Durham, Canterbury, York and Cambridge. The bridges were also noted, such as Magdalene Bridge, the earliest cast iron bridge in Cambridge in 1923, marking the original crossing point of the River Cam where the settlement grew up from Roman times. Lovell (ibid., p. 192) commented that the monumental architecture of gateways in cities such as Chester, Cambridge and Cantesbury are a frequent motif in tourist photography; a placemaker characterised Winchester as 'a city of gates'. Some of these gates could be argued to be decontextualised because they lack the original walls, which gave them their power as entry points in a stronghold. The gates retain symbolic resonance and are fragments of a vanished past, now positioned towards the centre of expanding historic cities, as opposed to marking their significant outer points of access. The gates provide an iconic, symbolic focus for the heritage gaze – for example, the nineteenth-century Brighton Pavilion gateway.

The awareness that 'original' historic cities are based on trade and other interests intersects with more recent time periods; for example, Lovell (2013) states how tourists observed educational motifs in Oxford and Cambridge, including bicycles, books and students. In contrast, Lincoln features an element of the more modern knowledge economy inherent in 'creative cities' (Florida, 2002; Landry, 2000), symbolised by a new university 'quarter' in renovated heritage buildings. This differs from Oxford and Cambridge, where the student places and activities are all-pervasive, making the Lincoln knowledge quarter seem more of an artificial construct. Lincoln has been segmented by the City Council into geographically based quarters, each of which appears to have a focal point – for example, Lincoln's 'cathedral' and 'cultural' quarters.

More distant periods of history are also exemplified by the might of monumental architecture and buildings of power, which are a key aspect of topography. Iconic, monumental structures, such as cathedrals or castles were, as Lynch (1972, p. 93) observes, long-term, radical, futuristic building projects which may have seemed hyperreal simulacra at the time and were certainly designed to create shock and awe. As Tuan (1977) details, many historic cities were designed to be defended against both nature and human enemies, with gates located at the four compass points and a circular shape created by the city walls. The parameter of the city's land was clearly defined; Thannington in Canterbury was originally named 'Thannington Without', referring to its location outside the city walls. City walls continue to reinforce their ancient function as fortifications and boundary markers enclosing the city.

The embodied tourist gaze (Urry and Larsen, 2011) gives physicality to heritage and results in the 'heritage effect' (MacDonald, 2009), where place influences behaviour. At pedestrian level, the monumental architecture of the city walls and their gates, as Tuan (1977, p. 162) argues, 'commands' and demands a specific emotional response. MacDonald (2009, p. 26) discusses architectural agency and the 'material relativities and material suggestions' which 'make a difference in the amount of effort people need to make to relate to them'. Discussing Nazi architecture, and quoting Speer's 'words in stone'. It could be argued that the 'architectural agency' of gigantic Nazi architecture is ideologically based, but the point is that the size has an impact on people, in the same way that Lovell (2013, p. 198) has described monumental cathedrals, city walls and winding streets of historic cities affecting tourists as immanent. The cathedral or a castle overlooking the skyline, provides the city with a panopticon (Foucault, 1995) for defence, and creates a visceral impression of visual dominance. The sense of authority is inherent, as Tuan (1977, p. 114) states: 'the building's centrality and commanding presence are immediately registered'. Most cities were, as Relph (1976 [2008], p. 71) observes, created by elite groups, and, in many instances, an abbey, castle or cathedral symbolises hegemonic power; for example, Durham and Rochester both have a predominant cathedral and castle, symbolising the battle between church and state. A Durham placemaker embraces the sensation of 'buildings of power' (Lovell, 2013, p. 193) in a muscular statement:

> Panorama. Sums up Durham. Cathedral and Castle. Designed and built together as symbol of Norman domination. Still standing, still dominant. Both cathedral and castle (now Uni) major players in city. Cathedral said to be finest example of Norman architecture in the world. No-one argues.

This comment is a remarkable insight into the original impact of the buildings, which marked historic cities from a distance by reputation as wealthy seats of power, compounding their aura. In Durham, and Rochester, the power struggle between church and state is architecturally embodied in the confrontational proximity of the cathedral and the castle. The cathedrals may have originally been intended to 'instruct', and inspire awe through the enclosure of space on a vast scale, but their continuing meaningfulness implies what Tuan (1977, p. 164) describes as 'enduring place', suggesting that monumental buildings have shared significance for both medieval and contemporary people, who may have a very different worldview: 'Any large feature in the landscape creates its own world, which may expand or contract with the passing concerns of the people, but which does not completely lose its identity.' These echo large structures all over the world, from the monuments of Washington in the USA to the Mayan temples and Ankor Wat, all of which are symbols of power.

Layout and centres

Lovell's (2013) research suggests that the layout of historic cities evokes a strong response; they can provide an unusual and exotic sight for the heritage gaze, where historic city street planning – non-grid, winding and narrow, with jetties – is imbued with the medieval pattern. Walking in a new environment with unpredictable street patterns may affect perceptions of historic cities: 'walking engages and shapes the world as envisaged, rather than merely seen' (Pink *et al.*, 2010, p. 6). People are conscious of the origin and impact of the medieval streets, lanes and alleyways which intersect. Lanes within historic cities are throwbacks to a rural past which might evoke feelings of a rural idyll and backstreets also present a backstage version of places, adding 'the chance of space' (Massey, 2005) evoked by the combination of people and place, the city fabric existing as a form of active interpretation.

Backstreets appear as unknown 'quantum authenticity' (Lovell, 2013, p. 200) and, as De Certeau (1984) describes, flow, appear and disappear unpredictably; for example, the backstreets and weeping willow by the river in Winchester, and Bath Place in Oxford. Lovell (ibid.) details how gardens seemingly contain 'hidden heritage' for local people, with backstage allure; they are closed, bounded by walls or gates, and exclusive; 'you have to know where to find them'. These backstage spaces are off the tourist trail and the pleasure lies in their strangeness, the sense of self-discovery, self-interpretation and the absence of other people.

Cathedrals seem to be the focal point of many historic cities, but the concept of places having a 'centre' is contended by Tuan (1977, p. 93) to be non-existent; he argues that 'anthropocentrism' is argued to be the human wish to perceive order. Lovell's (2013, p. 195) research demonstrated that anthropocentrism and anthropomorphism of place are evident in placemakers and tourists who search for and locate the 'heart' of the historic cities in a variety of streets and beside different landmarks. Placemakers identified 'meeting places', which have added personal and local significance. Like the 'bird's eye view of time', meeting places create encounters which take place on both a spatial and temporal basis, 'co-existing multiplicity' (Massey, 2005). Lovell (2013, p. 196) details how the meeting places and centres of historic cities were identified by tourists and placemakers as those locations which were inscribed with a heightened sense of the past activities and traces of workmanship and people – for example, the Buttermarket in Canterbury and Butter Cross in Winchester, both names clearly conveying their past use as dairy markets. A 'meeting place' in Lincoln is described as 'central to the High Street' (ibid., p. 197) and emerges as the 'actual, natural, centre' (in contrast to the cathedral) used by local people in an everyday capacity. Differing perceptions of city usage by tourists also create multiple centres. Canterbury's modern High Street and a wildflower meadow were both described as the 'heart' of the city. This implies that designing centres may not be effective; centres evolve out of the organic growth of cities and are where you perceive them to be. These are spaces endowed with personal and heritage meaning, where place and time intersect and different encounters can take place, leading to new possibilities, as Massey (2005, p. 62) details, 'out of each unique mixture new processes will arise'. Feeding the centres or 'hearts' of historic cities are arteries: the streets radiating outward dispersing people, or drawing people in towards the magnet attraction of the cathedral or castle. The central high streets and straight, Roman roads of Rochester, Chester, Durham and Lincoln form a river of time in contrast to the medieval lanes of Canterbury.

Ruins

The concept of 'genuineness' (Benjamin, 1936 [2008]) and 'originality' (Bruner, 1994, p. 401) are elusive in the modern world, where places are dynamic, continually eroded and renewed. Lynch (1972, p. 44) suggests that ruins provide an inherently 'original', historic environment and Lovell (2013) described them as real-real places. Their incompleteness also has visceral value. Kirschenblatt-Gimblett (1998, pp. 19–20) argues: 'The art of the metonym is an art that accepts the inherently fragmentary nature of an object. Showing it in all its partiality enhances the aura of its "realness".' Time collapses as we deconstruct and rebuild a structure like Fountains Abbey, near Ripon, in our mind, recreating what we know of the story and process; entropy ensures that the disintegrating nature of ruins adds to their sense of authenticity.

Ruination can be a constructed act; for example, in some de-restoration techniques (Otero-Pailos *et al.*, 2010, pp. 69–72) preservation is removed to create a stronger impression of integrity, as is discussed in Chapter 3. Allowing buildings to deteriorate is an act of what Cohen and Cohen (2012) term 'hot authentication', which can be loaded with ideological meaning: for example, the pilgrim shrine at Walsingham (Coleman, 2008), which was ravaged by the dissolution of the monasteries in 1533, and the site of the Nuremburg rallies, as MacDonald (2009) described, which was left to decay as a statement about fascism. Ruins can be used to invoke emotion; as Lynch (1972, p. 44) details, they impart a 'heightened sense of the flow of time' and the pleasurable sense of melancholy by reminding us that we are mortal and subject to time (Lynch, 1972; Wollheim, 1970; MacDonald, 2009).

A striking example of a ruin is the Brighton West Pier, which demonstrates liminal, evocative, nostalgic and melancholy qualities, as described by a placemaker (Lovell, 2013, p. 201):

> It's beautiful in its own way and a reminder of the way the seafront used to look. Visitors sit on the beach and stare at it as it holds some kind of magic: especially when the starlings in their hundreds form tight packed groups and formations flying around it at dusk.

This vivid description suggests a hot authentication of the pier. Other modern ruins which feature in Lovell's research include a rusting paddle steamer in the river Medway, which is a poignant sign of absent industrial heritage, and also the unoccupied Woolworth's storefront in Canterbury, a ghost of a commercial past, both of which articulate the process and power of entropy.

Brighton Pier, the Medway paddlesteamer and Woolworth's serve as metaphors for the passing of time and loss of industrial heritage: emotive versions of the 'bird's eye view of time'. Although placemakers may encourage perceptions of vibrant, modern places, as indicated by the findings in this chapter, visitors may be coming to historic cities to see evidence of the passing of time and appreciate the historic atmosphere.

Buildings with aspects of ruination include those visibly affected by time, as Lovell (ibid., p. 219) describes the 'moveable past' epitomised by the Old Chesil Rectory in Winchester, which has sunk a metre below street level, and also the crooked house in Lancaster. The King's School Shop in Canterbury is uneven and was described as being as if 'it could not be real'. Like the Leaning Tower of Pisa, these buildings convey a sense of time eroding space, the sense that permanence is unachievable and places are always becoming; the process of entropy is visible in glacial slowness. Describing Skiddaw in the Lake District, Massey (2005, p. 139) detailed how she realised that 'the mountains are moving, they simply move more slowly'.

Conclusion

The chapter has explored how the 'original' components of historic cities are not static; as Pred (1984, p. 279) articulates, places are always evolving; they are never 'finished', but always 'becoming'. This chapter has traced the journey of historic cities through a sequence of post-war events, beginning with the museumification involved in saving and rescuing architecture from modernism. Historic cities are perpetually being assembled; they are subject to changing ideologies, on display and consumed like spectacles by the heritage gaze, which means that modernisation is arguably more apparent. The heritage gaze is not passive, or limited to tourists, but includes subjective, performative acts of authentication by placemakers such as conservationists and planners, which make and remake the environment. Acts of authentication by those generating sources of official authenticity and those consuming it co-create the distance and mystery of aura. Entropy ensures that crumbling ruins, which are suggestive of the passing of time, provide the only inherently real-real, truly original historic environments.

The chapter suggested that the layout and built environment of heritage cities evoke a primary form of interpretation, incorporating affective elements, suggesting the inextricability of people and places. Deeply authentic, real-real experiences provide a bird's eye view of time, reinscribing the source of the city of origin's founding function under the surface, based on its location, trade or industrial past. Tourists authenticate historic cities by seeking their hearts, centres or meeting places, exploring the backstreets, discovering hidden heritage and personalising places. The tracing and retracing of the start, centre and purpose of historic cities, identifying what Benjamin (1936 [2008]) refers to as the 'fabric of tradition', is an instinctive need for many people, a way of reorientating themselves. This impulse to connect with the original city is a form of hot authentication, possibly driven by a desire to stop time, immerse the self and establish a sense of permanence in a changing world.

Historic cities encapsulate Bachelard's (1994) description of the poetics of the shell; historic cities are encompassed by a carapace of solid walls, the streets are winding and there is a murmur of the past, continually out of reach. Bachelard refers to the ammonite fossil, the shell turned to stone over time; the seemingly immutable buildings of historic cities create a sense of wonder that the history within them is ever-changing. Ammonites are curled in upon themselves – the spiral traced back to its origin at the centre – and the aura is vested in the inchoate mystery of their heft and ancientness.

3 The restoration and reproduction of historic sites

From the Ship of Theseus to mechanistic authentication

Introduction

More than any other, this chapter illustrates how places are singular, by their very nature; but sometimes they compromise this singularity by trying to be somewhere else and sometimes they succeed (see Figure 3.1).

Benjamin's (1936 [2008]) notion of aura has connotations of originality, rarity and singularity and any discussion of authenticity requires a full debate about reproduction and restoration. Heritage sites are globally highly significant tourist attractions (Benton, 2010; Fairclough *et al.*, 2008; Harrison, 1994, 2010, 2013; Harrison and Hitchcock, 2010; Jamal and Kim, 2005; Smith, 2006; Timothy, 2001) and, as illustrated by Historic England's 'Heritage Counts Report' (Historic England, 2015), heritage is not simply an economically significant force, but is central to the branding, identity and experience of many places. Historic environments may form an identifiably original 'heritage core' in many cities, but

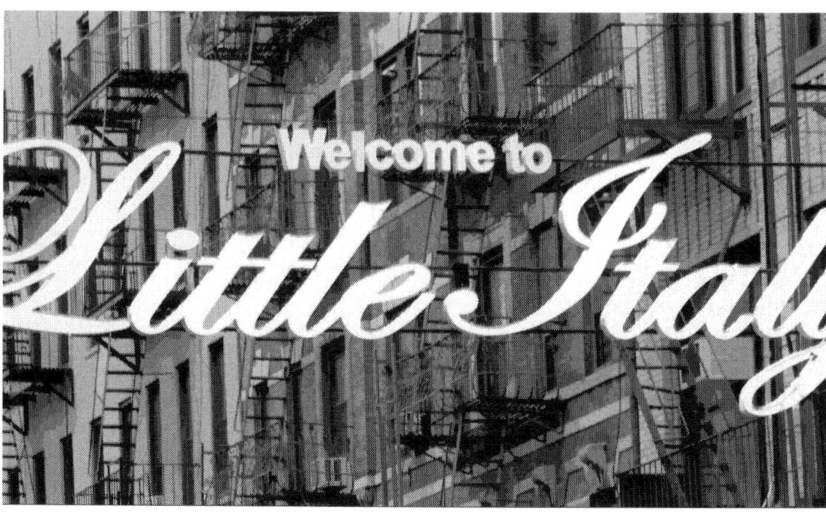

Figure 3.1 Little Italy sign, New York
Source: Tim Bolwell

they are also integral to most places and integrated into a vast number of locations, some of which are mentioned in this book, ranging from cultural destinations to seaside resorts. Urban heritage and heritage sites make insightful contexts for the assessment of different types of authenticity; while they are perishable if not preserved, new forms and models of cultural heritage are continuously being produced – thus, authenticity is always emerging, a process which is critiqued in detail in the chapter on gentrified places.

Heritage buildings are witnesses to the past, but their truth is far from unvarnished; instead, it is glossed, veneered and replaced. Heritage buildings were modern once: they were designed for a place and time and their originators did not picture the extent to which their creations would be reinvented in future years, nor did they imagine the modern settings which evolve around them. Heritage sites are timepieces: they are repaired, conserved, restored, adapted, embellished, decontextualised, recontextualised, recreated, replicated, destroyed, deconstructed and held together by a million tiny acts of authentication. Eco (1986, p. 16) described the 'reconstructive mania' which was symptomatic of hyperreality. All places are palimpsests, constantly evolving, while working to preserve an aura which, as stated by Benjamin (1936 [2008]), relies on a sense of time, distance and mystery.

The chapter examines acts of micro-authentication which restore and recreate heritage sites and the effects of these acts on tourists. The chapter uses Lovell's (2013) application of the combination forms of fake and real developed by Fjellman (1992), included examples of 'fake-real' restoration and 'real-fake' reproduction in heritage environments. The chapter introduces these typologies and the notion of the 'Ship of Theseus' (Plutarch), where objects such as timber ships have had all their component parts replaced, but are said to be the same object. Lovell (2013, p. 226) identified a cooler, fake-real state of mind, a 'mechanistic authentication' where tourists explore the workings of places and facts under the surface. As the chapter demonstrates, cool experiences may heat up, or even superheat to a state of wonder – for example, when contemplating the skills possessed by an unbroken line of craftsmen who restore buildings. The chapter also details the ways in which guided tours deconstruct heritage by providing information about copies. This can create a state of cool inauthentication, where the whole environment seems false. While historic environments are designed, authentication is participatory and tourists co-create the story of places. This chapter examines the actor-network in place (Harrison, 2013; van der Duim, 2007) viewing craftsmanship as a linking factor in a more-than-human world and suggesting that places are far from being in a state of post-authentication.

The movable historic environment

The reproduction and recreation of heritage sites is central to the evolution of places and the chapter identifies different types of reproduction, many of which are examined in detail in other chapters of this book. Replicas or 'authentic reproductions' (Bruner, 1994) could be said to be 'real fakes' (Fjellman, 1992; Hannigan, 1998) – copies which refer to original aspects of places. In urban environments recreations and replicas can be used as what Eco (1986) described as 'connective tissue', referring to the copies, models and dioramas intended to bridge gaps between artefacts in museums. Connective tissue is designed to enhance continuity, filling the 'spaces in between' buildings, avoiding the dissonance and decontextualisation created by the sudden juxtapositions of styles in a landscape.

Matisse's (1929) painting *The Treachery of Images* depicts a pipe, with the words 'this is not a pipe' written underneath. It is a representation of a pipe, not an actual pipe. In the same way, a copy of the painting is not the same as the original. It has not witnessed the same events, was not made by the same craftsman. It is a copy, but its aura often depends on whether we believe it is a copy, or whether there is a positive placebo effect because we believe it to be an original. Renewal can be described as a fake-real process because an innate part of maintaining heritage sites concerns turning back the process of aging. The constant repair of heritage sites is illustrated by a scene in the television programme, *Only Fools and Horses*. In the episode 'Heroes and Villains', a road-sweeper called Trigger explains to his friends that he has won an award for saving the Council money because he has had the same broom for 20 years. He then reveals that his broom has had 17 new heads and 14 new handles. When asked how it can be the

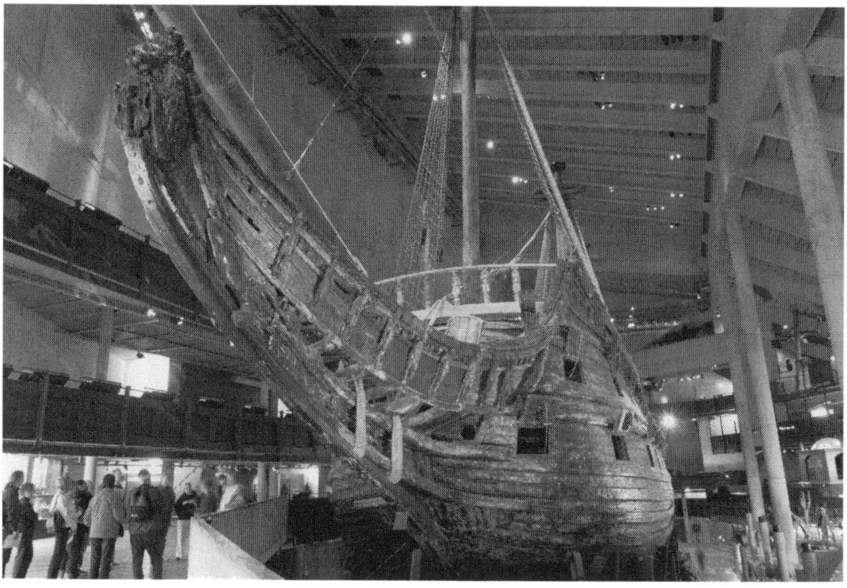

Figure 3.2 Vasa, Swedish warship with original wood preserved from the shipwreck
Source: Karolina Kristensson

same broom, he replies, 'Here's a picture of it, what more do you want?' Trigger's broom is a modern version of the theory of the Theseus paradox or Ship of Theseus, first recorded by Plutarch in *The Life of Theseus*, which examines one of the key questions about authenticity discussed in this chapter: whether an object is truly original and retains its aura if it has had all its component parts replaced.

On the one hand, people seek the singularity of an original such as the *Vasa* (see Figure 3.2) or *Mary Rose* warships, the original timber painstakingly preserved after they were raised from the seabed. On the other hand, a ship such as the *Cutty Sark* in Greenwich, which has had the majority of its timber replaced after deterioration through events such as fire, may convey the experience of a ship more effectively. Labadi (2010, p. 78) uses the term 'post-authenticity' when researching the mix of restoration and recreation included in World Heritage Site nomination dossiers, but heritage sites are far from post-authentication; heritage workmanship is practised at a granular level. The act of preservation is, in effect, artificial; as the chapter will show, restoration is an art which requires careful balance and it is important that the impact of preservation is also considered. Kirschenblatt-Gimblett (1998, p. 19) describes metonymy, where a fragment can act as a metaphor, allowing us to rebuild cities in our mind from their ruins, a process which is defined by Lakoff and Johnson (2008, p. 36) as 'the part for the whole'. Space is needed for our imaginations to roam in historic sites and sometimes we required the unvarnished truth.

Ironically, complete restoration may seem artificial in its totality, creating a sense of what Fitch (1990, p. 84) terms 'lost time', when our minds tell us that the place looks too pristine. Baudrillard (1981) contends that resurrecting the past is artificial; discussing the caves at Lascaux, he states the copies make both the real caves and the reproduction equally unreal. Historic sites are on display; Fitch (1990, p. 235) contests preservation as a 'Fourth Dimension', stating that 'the very act of exhibiting an artefact to public view – be it a painting, an old house, or an archaeological site – creates an unnatural interface between the viewer and the viewed'. The observation itself changes the experience and can't truly be apart from it; however, Fitch (ibid., p. 325) goes on to contend 'the need for alienated peoples to re-establish some experiential contact with the material evidence of their own past' echoing the value of the search for authenticity by tourists discussed in the introduction to this book.

Mechanistic authentication

Kirschenblatt-Gimblett (1998, p. 198) and Otero-Pailos *et al.* (2010, p. 74) described the 'heritage clock', where time moves on but the clock is constantly turned back. In J.K. Rowling's *Harry Potter and the Prisoner of Azkaban*, time-turners are devices that allow the main characters to go back in time, while trying not to change the past in a way that would disrupt the present day. Heritage conservation also strives to preserve the past as intact, without disruption. The concept of a 'heritage time-turner' encapsulates how we learn more about the past by turning time, through visiting sites, or consuming interpretation. In doing so, our perception of the past is altered and both the past and present are changed. Turning time allows us to see inside the heritage clock to its workings or mechanism (Heidegger, 1977, cited in Reisinger and Steiner, 2006). This has been further developed by Lovell (2013, p. 226) in a study of staged authenticity in historic cities as 'mechanistic authenticity'.

Studying tourist perceptions of staged authenticity in historic cities, Lovell (ibid.) identified mechanistic authenticity as a state emphasising the *accuracy* of historical knowledge; a factual, documentary approach to tourism, exploring the details – what McCarthy (2008, p. 536) terms the 'micro-histories' of objects. Certain academics have described hot and cool interpretation (Uzzell and Ballantyne, 1998), hot and cool authenticity (Selwyn, 1996) and hot and cool authentication (Cohen and Cohen, 2012). Mechanistic authentication is cooler, places act as prompts for information about their provenance. It is a state that is linked to Relph's (1976 [2008], pp. 51–4) 'objective outsideness', where elements of heritage space are viewed academically and impersonally. The 'cool heritage gaze' (adapted from Smith, 2006) could be said to be aware and active, resisting myth-making, or the construction of a tourist idyll, instead searching for factual authority and historical knowledge sources, which compound their value. Lovell (2013, p. 232) described how a sense of the appreciation of ancient skills is demonstrated by visitors to historic cities, who photograph and comment on craftsmanship in acts of micro-connoisseurship.

Cooler authenticity is also practised by placemakers. The UNESCO Nara Document on Authenticity (1994) includes a reference to 'wholeness and intactness… the physical fabric of the property and/or its significant features should be in good condition, and the impact of deterioration processes controlled'. The document considers the restoration and repair of historic sites, but places a strong emphasis on valid information sources detailing the materials and skills originally used. UNESCO guidelines state that 'reconstruction is acceptable only based on complete and detailed documentation and to no extent on conjecture' (2011). However, there is some relativity to the guidelines; the UNESCO document goes on to allow for authenticity of spirit and feeling: 'Attributes such as spirit and feeling do not lend themselves easily to practical applications of the conditions of authenticity, but nevertheless are important indicators of character and sense of place.' Asplet and Cooper (2000, cited in Chhabra, 2005, p. 65) observe the following criteria of objective authenticity, derived from the Aotearoa Maori Tourism Federation in New Zealand: 'from the mind of a Maori, by the hand of a Maori, and with a genealogical and spiritual connection to a tupuna Maori'. The mention of a 'spiritual connection' suggests the negotiable essence of the terms 'real' and 'fake' and it could be argued that the authentication of the historic environment is a constructed and social process, which is ultimately intangible and subjective.

As Lovell (2013) argues, mechanistic authentication may seem cool, but can be warmed, heated and superheated. Restoration can create a sense of 'heterochrony' (Foucault and Miskowiec, 1986, pp. 22–7) where several time periods are present as a 'co-existing multiplicity' (Massey, 2005) and comparisons can be drawn between them. The aesthetic gaze searches for residual signs of age and indicators of authenticity, such as patina and flaws (Lowenthal, 1985). Lowenthal (1998, p. 149) discusses 'the idea that old things should look old; faith in decay as a guarantor of antiquity; aesthetic delight in wear and tear'. Observing the passing of time is not simply intellectually engaging; the comparison between what was described by tourists as 'old and older' architecture becomes a source of wonder to some visitors, as demonstrated by this comment (Lovell, 2013, p. 227): 'Two types of bricks. A combination of flint stone and Roman brick. Both stonework of ancient time, but separated by one thousand years.' The observation supports Relph's (1976 [2008], pp. 51–4) category of 'behavioural insideness', where the visual aspects of places are studied by visitors for patterns and confirmation of knowledge. People search for the stories of places and these micro-histories add depth to their authentication.

Benjamin (1936 [2008]) writes of 'traces' of workmanship, not simply in terms of functionality, but also conveying the touch of the craftsman and, as Lovell's (2013, p. 251) research observed, when this is noted by tourists it can be highly affective. Conservation practices are mimetic; they involve ritual, a sense of the unbroken line of craftsmen; past skills are perpetuated and reproduced, their actions linking the past and present together. Crang's (1996, p. 421) work on re-enactments at a Tudor 'living museum' in Bradford indicated that, when taken to extremes, the technology and attention to detail in 'the material forms of authenticity' by re-enactors could almost be viewed as 'fetishisation', and the same could be said for the precision of the restorer. For example, in Greenwich the glazed, old-fashioned Victorian tiles along the British Foot Tunnel were replaced using expensive, crafted tiles from the Jackfield Museum in Ironbridge as part of the major Heritage Lottery Fund refurbishment work in preparation for the London Olympics in 2012. The perpetuation of crafts adds to the story, which reflects continuity and an impression of integrity in heritage sites.

Restoration

Labadi (2010) states that post-authenticity involves varying mixes of restoration and recreation; Lovell (2013, pp. 228–32) identified different types of reconstruction when researching staged authenticity in historic cities. The geography of authenticity includes combinations of many of these different aspects of restoration.

The first is where entire, invisible reconstructions take place – for example, the Campanile in Venice, which was completely rebuilt after it collapsed in 1902. The reconstructed building retains 'original' value for visitors unaware of its recent rebuilding, and the Campanile adds to the aura of the city by maintaining an iconic skyline. This approach has been used throughout historic cities – for example, Butchery Lane in Canterbury, where one side was rebuilt in 1992 and the other recreated after a more recent fire, but both seem as if they are the same age and fit with the city, being situated near timbered buildings dating from 1450 – suggesting that restoration can seem seamless.

Large-scale reconstruction in historic cities, such as the old centres of cities like Gdansk, Warsaw, or Dresden, which were bombed during recent wars and meticulously restored in the early twenty-first century, present another type of restoration. The difference here is that these structures were restored many years after the original destruction took place and they contain an ideological element of collective commemoration. The project can bring feelings of growth, optimism and what Message (2006) describes as 'making whole'. For tourists who are uninformed that the buildings were destroyed at all, seamless copies may pass as well-maintained originals and convey charm and wonder. Gdansk drew data from accurate information sources as specified in the Nara Document on Authenticity (2007) and was aiming for 'verisimilitude', a state which Bruner (1994, p. 399) has defined as where the copy of the original is flawless. Relph (1976 [2008], p. 68) argues that places based on nostalgia are self-conscious and inauthentic and Fitch (1990) embraces the idea that extensive reconstruction is 'fictitious' involving 'concealing' changes to make them 'look like new'. Huxtable (1997, p. 18) takes the position that this type of copy lacks depth, suggesting that only the original has value and aura, 'To equate a replica with the genuine artefact is the height of sophistry… what is missing is the original mind, hand, material and eye.' However, the sense of wonder at the reconstruction of cities such as Dresden and Dubrovnik can generate feelings of ideologically hot authentication vested in the significance of the act of rebuilding. These are classified as first-order simulacra (Baudrillard, 1981) in the sense that they are faithfully executed copies of the original.

The theory of the Ship of Theseus encompasses a further type of restoration, involving the preservation and renewal of perishable materials. The Stave Churches in Norway are restored, repaired, large sections are original and they have been rebuilt after fire damage (see Figure 3.3). Japanese temples and castles such as Himeji, or the wooden statues at the shrine of Nikko Toshogu require 'invisible restorations', where the impression is historic but the mind knows that this is not the original material and thus accepts the restoration and craftsmanship as implicit. This type of restoration could be said to include structures such as HMS *Victory*, where components have been replaced to the extent that very little remains of the original, yet the item still retains its identity and has evocative power as recognisable object.

Visible craftsmanship is an important contributor to the authentication of historic sites, revealing the mechanism of historic structures and the visible process of change, what Jordan (2006) describes as 'structures of memory'. To the untrained eye, preservation may be seamless and invisible, without overly embellishing or stating the obvious; verisimilitude keeps the illusion of truth-likeness intact. The reconstruction is false (hence fake-real), but its intention is nearer to the truth. Craftsmanship can be a form of practiced mimesis and some heritage buildings, such as cathedrals, reflect the different craftsmen who worked on them, who were often modelled in the gargoyles and carvings. The nuance differentiating this type of restoration is that the historic environment is rarely restored in its entirety; instead, the process is continuous. Lovell (2013,

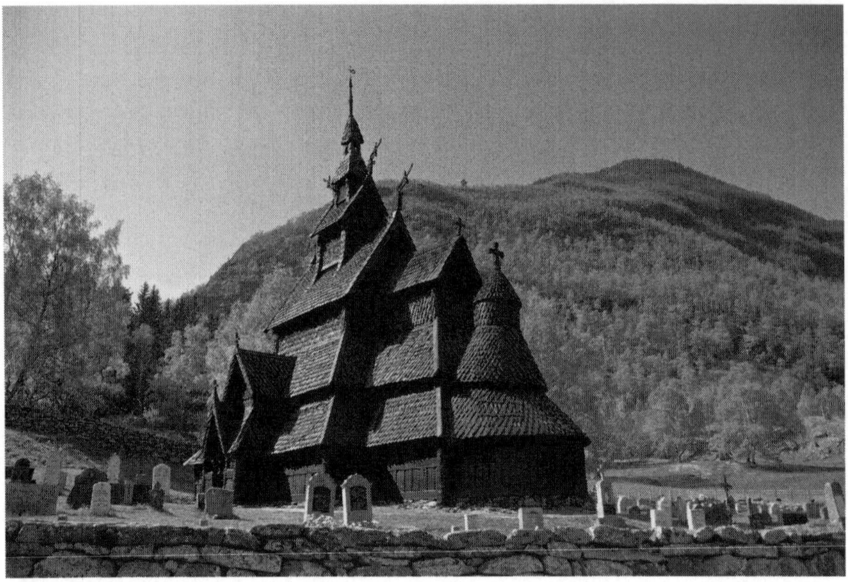

Figure 3.3 Borgund Stave Church
Source: Visit Norway, www.visitNorway.com

p. 228) details how, where skills are specialised, craftsmanship is recognised by visitors as a rare and anachronistic activity and a source of wonder. Tourists observed the craftsmanship of Rose window in York, maintained by the York Glaziers Trust, involving intricate, stained-glass restoration skills.

De-restoration techniques (Otero-Pailos *et al.*, 2010, pp. 69–72) add another element to the mix. Repair and conservation can make stonemasonry seem cleaner, whereas stained glass renovation allows the joins to show. Lovell (2013, p. 232) stated how ancient documents in Canterbury Cathedral Archives require visible repair. De-restoration involves either removing preservation, or making it visible: for example, pottery fragments which are pieced together with missing aspects presented in a different colour to make them apparent. The difference between restored and 'authentic' pieces in Canterbury Cathedral gate can be determined by the hue of age. However, visible preservation can give the effect of being poorly done, which can undermine place authenticity with what Fitch (1990, p. 84) argues to be the 'local (and often amateurish) concepts of "making it look the way it really was"'. Museums will often place a deteriorated object and a replica side by side to illustrate the processes of time. One of the authors observed how many visitors admired and photographed the recreation of the Sutton Hoo helmet in the British Museum more than the original helmet. Sometimes the varnished truth is more truth-like than the unvarnished.

Interpretation, usually seen in museum environments, is a distinctive act of restoration, placing what Samuel (1994, p. 129) terms the 'rescue' narrative at the centre of the display. The 37-metre-long Viking warship in the British Museum Viking exhibition in 2014, the *Mary Rose* Tudor warship in Portsmouth, the Swedish warship *Vasa* in the Vasa Museum in Stockholm and the *Cutty Sark* clipper in Greenwich are maintained in a way that allows visitors to appreciate their scale. Their conservation supplies a 'back story' (Vogler, 1999, p. 94), the interpretation focuses on the salvage of the wrecks, and the treatment and maintenance of the perishable timbers plays a key part in the narrative, the 'Fourth Dimension' to which Fitch (1990, p. 325) referred in the introduction.

Gazing not only behind the scenes, but above and below them also reveals hidden restoration. Visible restoration can provide backstage glimpses (Goffman, 1959; MacCannell, 1973; Moscardo and Pearce, 1986; Edensor, 2001) behind the scenes of personnel at work. Many cathedrals and other heritage buildings maintain their exclusivity through a 'restricted view' learning experience, providing a 'staged backstage' – for example, through tours or specific events, such as the Stonemason's Festival at Canterbury Cathedral, or the cleaning of the gargoyles at Rouen Cathedral. The restoration of Canterbury Cathedral's roof leading is carried out in the open in the cathedral precincts, which adds significance because the roof is above the eye-line, so visitors gain a close-up experience of rarely seen detail above, rather than behind, the scenes.

Seeing 'below the scenes' is evident in extensive archaeological work, which may be necessary during the development of historic sites and which provides additional authentication of their provenance, offering a glimpse 'under the stage' to some hidden treasures. Archaeology rarely fully reconstructs what used to stand, except in digital or virtual reality. Some archaeological sites may not seem immediately entertaining for visitors, so when the Whitefriars Shopping Development was built in Canterbury in 2005, Canterbury Archaeological Trust conducted 'The Big Dig' on the site. The extensive dig revealed the original medieval street pattern of the city, which was then recreated through the layout of the shopping centre buildings. The Big Dig provided rich material for staging; it contained a temporary shelter displaying found objects and a walkway over the site where visitors could see the work and the tracings of heritage buildings. Similar archaeological work took place during the construction of the 1980s Coppergate development in York. Both archaeological digs created on-site commemorations of social memory: Coppergate now includes the Jorvik Centre, which displays the finds of the dig and celebrates Viking history; in Canterbury a permanent record of the archaeology was created by an artist in residence at The Big Dig. The artist used the archaeological blueprints of the dig, which recorded where a variety of medieval artefacts, such as combs, were found. The images of the artefacts and blueprints were traced by the artist into the paved street surface, which therefore contains impressions of the past, for the future. There is no guide to illuminate the artwork, which can appear abstract underfoot, but to a place-maker who understood its meaning it had added significance as a memorial and reminder in Lovell's study (2013, p. 284), stating: 'we are but custodians of an historic environment'. Restoration may thus lead to authentication and a sense of custodianship, which is essential in the epoch of austerity measures and funding cuts in heritage maintenance.

Types of reproduction and recreation

Eco (1986) relates how the relationship between 'reproductions of reproductions' and original artefacts in museums is blurred and irrelevant, where some fabricated objects serve as 'connective tissue' to reinforce the overall effect. When restoration is applied on a city-wide basis, reproductions and recreations, or real-fake buildings, bridge the gaps between historic buildings, as is explored in the chapter on gentrification. Postmodern and pastiche styles use references from the past and poetic licence to enliven contemporary objects, reducing Fitch's (1990) sense of 'lost time' by closing the gap between the modern day and remote periods of history through the story of the journey – for example, the new shopping centre in Bath which uses local materials. Real-fake buildings can seem nostalgia-driven, as new developments are embedded and rooted using mimetic design references to the past; replication is one of the key features of postmodern design and global sense of place (Massey, 2005). Reproductions are not intended to act as a focal point, or interrupt the heritage experience and sense of place; they articulate the spaces in between, which are intended to be invisible, locating places using vernacular references. Real-fake buildings are intended to create continuity, avoiding the architectural dissonance of historic buildings in overly modern environments.

There is a difference between an exact copy in situ, a *restoration* of an original, such as the buildings of Dresden, a *reproduction* and a *recreation* that is modelled on an original. Reproductions (fake-reals) are faithful copies, the intention is that they are truth-like; complete replication is an impossibility, because they are made at different times from the originals on which they are modelled. Recreations (real-fakes) are more imaginative and inventive, emulating the spirit and feeling of the original, which is reflected in the ways in which they are authenticated by tourists. Benjamin (1936 [2008], p. 11) describes objects as having an 'original, initial, utility value', and the function of copies is different. Bruner (1994, p. 399) uses the term 'authentic reproduction' in his discussion of the recreated New Salem, to denote a convincing reproduction of an original. Bruner's phrase is critiqued by Huxtable (1997, p. 18): 'Authentic is the real thing, and a reproduction, by definition, is not; a copy is still a copy, no matter how skilled or earnest its intentions.' Lovell (2013) suggests that real-fakes may create an impression of authenticity or sense of place by referencing a historical period and striving to convey the spirit of the original. Stylistically, some new buildings in historic cities can be described as 'contemporary contextual' (ibid., p. 269) using the vernacular to supplement what Meethan (1996) terms 'spatial narratives'. Contemporary contextual builds are not necessarily designed to act as aesthetic focal points, but as locally distinctive settings for the historic environment. Relph (1976 [2008]) suggests that self-consciously created places are based on nostalgia and are therefore inauthentic, but it could be argued that new building developments which reference past styles add value to the holistic integrity of historic places. The chapter on gentrification debates the influence and value of pastiche in detail.

There are many internationally recognisable forms of historic reproduction (although the discernment of inauthenticity may depend on familiarity with the subject, local or expert knowledge). The following examples are by no means exhaustive, but they serve to illustrate the infinite combinations of fake, real and hyperreal in a range of heritage imitations. They are not placed on a scale of authenticity because, as the book argues, authenticity is a fluid, permeable concept.

The first example involves the copies of iconic buildings that function as part of the architectural landscape. In particular, Italian stylisation may add a double layer of historical reference – for example, the 'Bridge of Sighs', Hertford Bridge, at the entrance to New College Lane in Oxford, which resembles the Rialto Bridge in Venice. This glimpse of Italy complements the historic city (there is also a Bridge of Sighs in Cambridge) and does not seem out of place because, although the bridge is authentic to Venice, it brings a nod of 'Veniceness' to Oxford. Because of the classical context of Oxford, the iconic bridge seems less derivative and is accepted, although it is a copy from the city of Venice. Certain other architectural structures – for example, the domes of university colleges and libraries – epitomise the recognition that Oxford is part of a joined-up classical world. Italian influences were likely to have been brought back by travelling religious (and scholarly) inhabitants of UK historic cities and consumers of the environment are sensitive to the sources of stylistic derivation. The Bridge of Sighs, while exquisitely crafted, is still an intentional copy, providing a different experience. It is valued as an Oxford icon and is therefore a mixture of two myths: Oxford-Veniceness. Of course, it must be noted that Venice borrowed heavily from elsewhere; the Bascilica di San Marco façade incorporates a hodgepodge of artefacts brought back from trading partners, such as horses from Constantinople and the oriental pillars on the portico. As the introductory chapter states, while places are singular by their very nature, often they try to be somewhere else. Examples of this are Chinatown and Little Italy (see Figure 3.1) in New York. Interestingly, urban villages such as Chinatown resembled a fusion of New York-Chinaness authentication (see Figure 3.4), with New York's distinctive architecture and the smells, products and foods of China.

While this book does not explore marketing in detail, examples of places consciously copying other places include the 'Venices of the North' such as Bruges in Belgium, Wroclaw in Poland, and Birmingham and Little Venice in Maida Vale in the UK. The two 'Venices of the East' include Udaipur in India and Suzhou in China, while Basra is said to be the 'Venice of the Middle East', Recife is the 'Venice of Brazil' and Annecy the 'Venice of France' (*Telegraph*, 2016).

Other 'Venices' (or Other Venices, designed for the 'tourist Other') include Dalian in the north-eastern Liaoning province of China, which closely references Venice's canals and architecture, as does Baku in Azerbaijan, which replicates features of Venice. Rescaled copies of Venice, in Las Vegas (see Figure 3.5) and Epcott, gain their own historicity, they are examples of second-order simulacra in jokes, which rely on knowingness.

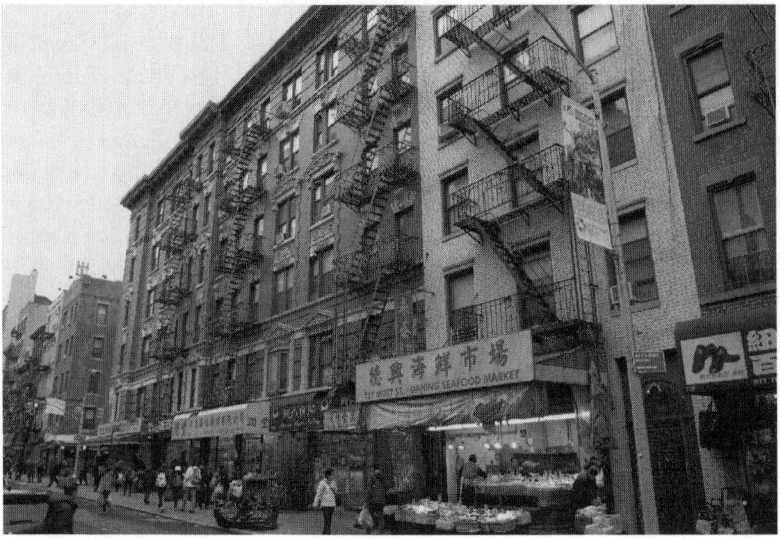

Figure 3.4 Chinatown fire escapes, New York
Source: Tim Bolwell

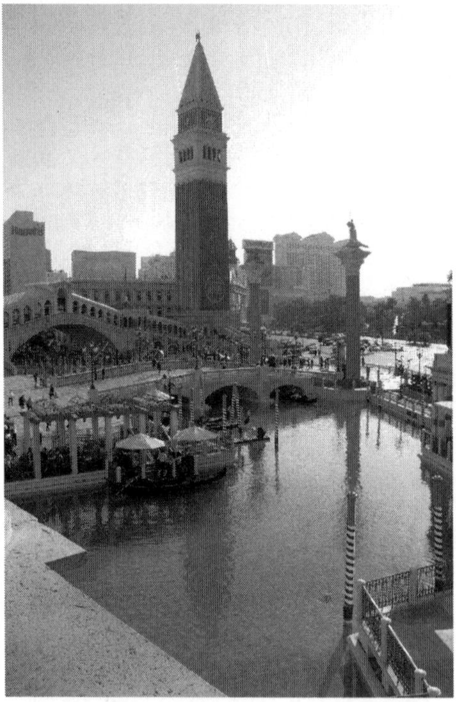

Figure 3.5 Venice, Las Vegas
Source: Author

The replicas continue, like Russian dolls: in Legoland Germany there is a copy of Venice, which allows visitors to feel like giants as they view the miniature St Mark's Square and Doge's Palace. They illuminate a further permutation of reproduction, which involves 'inauthentic reproductions' (adapted from Bruner, 1994), the decontextualised use of life-size and scale models of iconic architectural buildings and destinations for theming. Some constructive authenticity theories are based on ideas of dislocation (Bruner, 1994; Delyser, 1999; Fjellman, 1992). The clearest example of this phenomenon is stated by Salamone (1997), who compares two San Angel inns in Mexico City and Disney World, managed by the same proprietors. Both restaurants retained a contextual Mexican authenticity, contingent on 'a relational process, rather than an object' (ibid., p. 308) and, while the Mexican-located restaurant was European in atmosphere, the Walt Disney World version included Mexican caricatures, providing a surrogate Mexican experience.

There are several replicated places in China – for example, the Eiffel Tower in Tianducheng, near Shanghai. Wainwright (2013) uses various sources to describe this concept:

> Sima Qian, who recorded a remarkable building programme pursued by China's first ruling dynasty, the Qin: 'Whenever Qin conquered one of its rivals, it would commission replicas of its palaces and halls and reconstruct them on the slope north of the capital. From Yongmen all the way to the Jing and Wei rivers, there were replica palaces, passages, and walled pavilions.'

Sam Jacob also adds that 'if you look at early buildings by the architectural firm SOM (Skidmore, Owings and Merrill), they are indistinguishable from Mies van der Rohe; so much so, in fact, that Frank Lloyd Wright nicknamed the trio The Three Blind Mies' (cited in ibid.). Both Las Vegas and Disneyland place different eras and worlds together in a similar proximity to the Cast Courts at the Victoria and Albert Museum, which were provided as a Victorian method of surrogate travel and source of wonder, plaster copies of Trajan's Column juxtaposed with the façade of Santiago di Compostela and Michelangelo's *David.* Wainwright (2013) cites Jacob, 'whose Museum of Copying appeared in the 2012 Venice Biennale, containing a replica of Palladio's Villa Rotunda in spray-foam and plaster. "Copying antiquity is what you were supposed to do, and new things emerged."' These examples may seem to challenge the singularity of place; they attempt to bring places closer. However, contextualisation alters the essence of places; the Eiffel Tower is not simply 'the Eiffel Tower' if it is not in Paris, it becomes the 'Eiffel Tower in Vegas', or the 'Eiffel Tower in Tianducheng'. The replicas are thus distinguished from the original; a picture of a pipe is not a pipe.

Hidden copies provide another form of reproduction; they have added significance because they may result in the 'heritage placebo effect' (Lovell and Beedie, 2011, cited in Lovell, 2013, p. 281). An example of this phenomenon is apparent in St Michaels Arcade at Bridge Street in Chester, which was finished in marble circa 1902, until a public outcry led to a black and white Elizabethan façade. Reproducing the black and white timber, which was characteristic of Chester, gives a general 'Tudor feel', although the façades are Victorian. The positive placebo effect lay in the belief that the buildings were Tudor, and disillusionment on finding they were copies. A realisation that tourists have had 'an authentic experience of a copy' may make them reframe their experience, as will be discussed further later in the chapter. Other examples of positive heritage placebo effects include the ducking stool on the river in Canterbury, which is a Victorian, rather than a medieval addition.

Reproductions derived from historical documents or sources are also significant. The intention to create a medieval herbarium in Canterbury Cathedral was stated by a cathedral employee:

> We know that there was an herb garden, a herbarium, at the medieval cathedral but its layout is only indicated in one drawing. So it was possible to be creative when designing the modern garden, as it wasn't possible to be authentic.
>
> Ibid.

The cathedral had followed the UNESCO guidance on spirit and feeling and their vision of the past was a 'reinvented tradition' (adapted from Hobsbawm and Ranger, 1983) adding to the overall atmosphere of the cathedral precincts. This also created a positive heritage placebo effect for visitors – the herbarium resonated with those who believed it intact and perpetuated, but it was recently constructed by the site of what was the monks' toilet block, as revealed on the official Precincts tour (Lovell, 2013, p. 281). Examples of other recreated historic gardens include the Knot Garden at Mosely Old Hall and the scented garden at Kenilworth Castle, but, in these cases, visitors are aware that these gardens are sequels, taken from original sources and involving the retracing of older steps.

Some examples of reproduction serve the function of protecting the original exhibit – for example, the replica Lascaux Grotto, opened in 2016, which was deemed to be necessary because the breath of visitors was damaging the original Cro-Magnon cave paintings. As described by Willsher (2016):

> Stepping into the replica grotto, built 800 metres down the hill from the original with the same sombre, humid atmosphere, visitors are greeted with paintings of herds of animals, overlaid on earlier older paintings or etched into stone, that have taken a dedicated team of modellers, sculptors and artists three years to reproduce, as near to the Cro-Magnon original as possible… Archaeologist Jean-Pierre Chadelle described it as a work of art in itself and said work on the replica had enabled experts to discover new details about the original Lascaux.

This comment illustrates how constructing recreations involves skills of authentication which can be as intricate as the minutiae involved in restoring original buildings and artefacts. A further example is the copy of the statue of Michelangelo's *David*, in the Piazza della Signoria in Florence, in 1910. The original is now in the Accademia, but the copy, placed in the same position as the original, is arguably the most-photographed and supersedes the original as locationally iconic. As Lovell (2013) discussed, there is further complexity in the representational value of statues, which depict a 'real' person.

In a similar vein, another form of reproduction involves the simulated, hyperreal recreation of the whole, through digital and virtual effects, headsets, apps and other means. This can also take the form of events such as the Cathédrales de Lumière light show, staged by one of the authors and designed to animate while protecting the fabric of heritage buildings. The event linked the cathedrals of Amiens, Rouen, Rochester and Canterbury, recreating the 'painted past' (Lovell, 2013) on the portico of Amiens cathedral. Amiens was not subject to the Victorian cleaners who had scoured many English cathedrals, and the polychromatic paint fragments in the French cathedral enabled the event to more accurately establish a colourful restoration of how the cathedral would have looked, which, to modern eyes, was almost gaudily inauthentic; the saints had rosy cheeks and the clothes were bright and patterned. These recreations are protective, as in the example of the cathedrals, but they are also about extending the effect and allowing heterochrony, enabling the viewer to travel in time to historical events – for example, the fire of London – illustrated in the game Minecraft.

Other types of recreation and reproduction are discussed in detail elsewhere in the book, but will be mentioned here in brief. Faithful reproductions – for example, historical theme parks such as Colonial Williamsburg – are mentioned in the chapter on theming, as are fictional places created by imagineered authenticity, such as theme parks like The Wizarding World of Harry Potter. Chapter 7, on the rural idyll, discusses the parks and gardens associated with great country houses and stately homes planned in the style of paintings, using Arcadian principles. Those places and simulacra recreated from the artistic dreamscapes are illustrated in Chapter 5, on art. Conversely, regional places, which are given fictional cloaks of induced identity such as 'Robin Hood country', are also addressed in Chapter 11. The value of historic re-enactment and living history is discussed in Chapter 10, on events, and Chapter 4, on gentrified places, assesses the way in which industrial heritage sites are revived and repurposed with pastiche design references. Chapter 9, about filmic places, assesses images and representations of historic places in photographs and on postcards and souvenirs, as well as in social media memes and gifs. These are only some of many permutations, the list is endless – fake and real twisting in different incarnations, like DNA.

Some places are neither completely reproduction nor recreation, but have an air of unreality about them. Ideologically constructed places that are drawn from utopic plans have an air of restoration – for example, model villages constructed by Victorian industrialists and Quakers, such as Saltaire, built by Sir Titus Salt, or Port Sunlight, constructed by William Lever for his workers and Prince Charles's Poundbury, which uses neo-classical and reproductive styles. Some places always retain a sense of being 'concept cities', even when inhabited. New garden towns and villages may initially have a sense of unreality about them, but not to the extent of model villages, which retain the aura of their Quaker values and are also on display as industrial heritage tourist attractions.

It is important to note that the processes of recreation and restoration are becoming increasingly transparent. One of the more recent aspects of restoration, discussed further in Chapters 4 and 5, is the use of cultural interventions and digital placemaking to transform and reanimate heritage places through creative interpretation. In the climate of funding cuts, many museums have become trusts and volunteering culture now plays a fundamental role, both in the maintenance of many historic places and in maintaining organisations such as the National Trust and English Heritage. Creativc interpretation employs participative techniques to play with the spirit and feeling of places, engage and immerse people, discover fresh perspectives and find different ways of authenticating heritage sites. Creative authentication invests volunteers with a sense of custodianship, ownership and 'creative stewardship' (Lovell and Brown, 2014) which is essential in times of austerity or change.

Official authentication

Another form of mechanistic authentication involves access to the facts behind historic sites. 'Authority' was one of the four categories developed by Bruner (1994, p. 401) to discuss the authentication of objects and events. Authority is evident in plaques, interpretation, World Heritage Site designation and through verification by experts, who may also subscribe to the 'authorised heritage discourse' (AHD) (Smith, 2006). Official authenticity merges space and time by locating the past events of the historic sites and superimposing them onto the present-day physical landscape. For example, Lovell (2013) explored how people were interested in the 'firsts' that historic cities had to offer, such as the Ashmolean Museum in Oxford, the world's first public museum.

Historic plaques initially seem to be a stamp of authority or official authentication, but, as Jordan (2006) established in her study of Berlin, plaques may be randomly created by many different heritage bodies over decades. Forms of 'casual interpretation' (Goodey, 2006) can occur where a heritage place has been demolished, or disappeared, leaving a decontextualised plaque or marker, usually at ground level. These markers can seem to possess agency by supplying a connection to past events (Loewen, 1999, p. 27) and providing a surrogate experience (Uzzell and Ballantyne, 1998, p. 233). Plaques are part of the 'constellation of names' (De Certeau, 1988, p. 104) discovered in a serendipitous, situationist manner by independent travellers, or pointed out to groups by their tour guide. Plaques are seen as distinguishing authentic from inauthentic heritage by tourists in historic cities, as demonstrated by this comment in Lovell's (2013) study: '"Real or fake?" That was the question I asked my husband while looking at the building in this photo. He replied "It must be real; it has a plaque on it!"'

Guided walking tours are key forms of collective memory-making and 'remembering together' (Smith, 2006), offering an officially authenticated, 'narrativised' (Shudson, 1995, cited in Benton and Cecil, 2010, p. 18) version of the social history of historic cities. Tours are choreographed, multi-sensual, 'directed performances' (Edensor, 2001) of heritage tourism: involving moving, looking, listening, reflecting and photographing. This is the opposite of Lovell's (2013) 'quantum city' of self-discovery; the guided tours seem to give backstage access to information about heritage sites, but their choreography shapes the selective knowledge which is imparted. Reisinger and Steiner discuss (2006, p. 482) how guides provide an edited 'script' which is filtered through Blue Badge training, encouraging adoption of the authorised heritage discourse and shaping places through decisions such as the selection of the highlights. It could be said that these backstage versions of historic sites created by tours offer restricted views, steering tourists in directions that reflect the guiding organisations' own field of expertise and micro-histories.

Guided heritage interpretation involves an information flow, largely in one direction, from an 'official' expert to a non-expert, sometimes reframing an interpretation of a place or event. Guided tours are also conduits to collective authentication; the tour group can be assumed to share social memories in a two-way process. In addition, participants can resist tours and it should not be assumed that tour participants are passive (Urry and Larsen, 2011; MacDonald, 2009). Apocryphal stories indicate that guides narrativise memory by interpreting, spicing up and subverting the authorised heritage discourse. Lovell (2013) observed how the guides have themselves become valued heritage artefacts by tourists, repositories of collective memory and deemed authoritative because of variables such as age, profession and residency. The authority of the guides as authenticators of place is visible in their blue badge or yellow sash, emphasising the validity of the information they impart. Crang's (1996) research discussed how re-enactors take details of authenticity seriously, as do tour guides through their training.

Adaptations may be invisible to the uninformed 'heritage gaze', but experts possess detailed knowledge about their domain which unveils the story behind the façade. For example, a painted, much-photographed house in Palace Street in Canterbury is an 'old and older' assemblage of historic styles, as illustrated by the following tourist comment (Lovell, 2013, p. 238): 'Painted House Northgate. All is not what it might seem. Behind the painted frontage lies a Norman stone house. The "new façade" was put on in the 17th century lifted from another building in Kent. Dissembling is not purely a modern art form!'

Lovell observed how, when tourists went on guided tours and assimilated insights or facts such as those detailed here, their reframed knowledge could render seemingly original buildings inauthentic, leading to a state of 'cool inauthentication'. The aura of the building was then also shifted to the mechanistic fact. This seemed to be the case even if the addition was made hundreds of years before and is therefore a historic adaptation, as illustrated by this tourist comment (ibid.): 'Conquest House. This house really looks old. The guide tells me that it is a quality fake house front. It certainly is! Thought it was authentic.' Detailed knowledge can offer new reappraisals of buildings that tourists had perceived as authentic, as illustrated here, 'Thomas Beckett shrine added onto cathedral as tourist (pilgrim) attraction therefore inauthentic' (ibid.).

The use of eighteenth century mathematical tiles, which were made to look like bricks and battened onto the frontage of buildings in Canterbury, is an example of a micro-history within guided tours. Mathematical tiles are considered a feature of local distinctiveness and guides are advised to include them in their scripts, making them part of the authorised heritage discourse. When tourists are shown the tiles, and become aware of the mechanism of the 'heritage time-turner', the integrity of place is disrupted, causing dissonance, a state observed by two visitors who have been sensitised to the façading technique by the Canterbury city tour (ibid., p. 239): 'After all the stories of "fake/new" front[s] on buildings you start to wonder what is real and what isn't.... Walls around Canterbury Castle. This city wall has so many rebuilds – which is 'real'? Which part is Roman?'

The tourist observation suggests a change in attitude from an acceptance of the historic appearance of the city on the surface, to a reframed, deeper questioning which does not reflect tolerance of post-truth or post-authenticity. 'Mechanistic authentication' could therefore be argued to block the suspension of disbelief necessary to appreciate deep history, which may be more apparent during independent self-exploration. It can lead to 'cool inauthentication', when people gaze at historic sites, wondering what else is 'inauthentic'.

The authorised heritage discourse involves cool authentication, which is 'cognitive and epistemological' (Uzzell and Ballantyne, 1998, p. 152). The past does not largely seem to be framed in a contested way by the guides; instead, they can be 'conformist' (Reisinger and Steiner, 2006, p. 482). While some tourists on tours appreciate learning an official perspective about how historic places are renewed, others find it provokes deeper questioning about the nature of history. However, as Lovell (2013, p. 237) details, the apocryphal stories indicate how guides narrativise memory by interpreting, spicing up and subverting the authorised heritage discourse.

The knowledge provided by guided tours can also reframe the gaze from the modern time period to the past, as related by the following comment in Lovell's research (ibid., p. 240):

> Queen Elizabeth's Guest Chamber/Cafe Nero's. Again, new and old combined. I was passing it yesterday and decided to take a picture because of the horrific sign. Today we stopped and the guide told us about the building and the story behind it. Standing there watching it, the building won and therefore I took a photo.

The comment demonstrates how the tour enriched a tourist's view of a historic building, restoring its aura and stimulating the imagination, rather than rupturing the moment. It is an extremely significant comment because it clearly demonstrates temporal flow, showing how authenticity emerges and retreats.

Conclusion

This chapter has examined the practice and consumption of cooler, constructed and more cognitive fake-real forms of authentication in relation to heritage sites. Like the Ship of Theseus, places are continuously made and remade, the time-turner set and reset, collective memories reinscribed and new experiences created, with implications for the originality of places.

The chapter discusses mechanistic authentication, outlining different types of fake-real restoration: reconstruction; ideological large-scale reconstruction many years after the date, with an ideology grounded in commemoration; the restoration of entire structures made from perishable materials; preservation using original skills; de-restoration and the unvarnished truth, where the rescue narrative supersedes the restored object and restoration, not simply backstage, but under and above the stage. The chapter also discusses types of real-fake reproduction, assessed both here and encountered elsewhere within this book, including copies of iconic buildings, rescaled replicas which are decontextualised, secret reproductions, recreations from historical or fictional sources, simulated recreation and representation as images. Mixing styles and adapting the uses of historic buildings reduces 'lost time' (Fitch, 1990) and closes the distance between the modern day and remote periods of history. Restoration of the historic environment is shown to be on display, making it a form of staged authenticity, an eclectic array and a living canvas where the visitor can fill in the gaps, deconstructing and rebuilding places. In other chapters, the aura of places is encountered; in this chapter it is excavated: perhaps rather than '*seeing into*' (Relph, 1976 [2008]) heritage places, the mechanistic authentic gaze involves *seeing* historic sites and *reflecting on* them.

Official interpretation appears to underpin the museumification of historic places by objectifying them. Official authenticity allows both experts and tourists to use knowledge to add distance to their experience of historic sites, which is a cognitive method of enhancing value and compounding aura. An examination of guided tours demonstrates that the heritage time-turner can change the way tourists view the present. However, insider information provided by guides may make the cool heritage gaze both more critical, ultimately inducing a state of cool inauthentication, where tourists wonder about the extent of artifice in heritage places. Collective authentication can also superheat the gaze, taking people back into the past and hotly authenticating places. Mechanistic authentication is comparative and explores the age of the historic environment as a source of wonder, an aesthetic response to time.

The slow, authenticating gaze takes in the panorama of heritage sites, from their topography to their patina, all which add extra layers of clues about their journey and its meaning. It is important to note that tourists don't simply consume the surface veneers of places, but also the assembled acts of conservation, restoration and preservation – both in caring for original structures and assembling replicas, which give historic sites resonance. Historic places are held together by a million craft repairs, providing continuity, integrity and artistry in nails, wood, glass, stone, string, straw, wattle and daub. They are also held together by expectations created by the superstructure of heritage. The truth is rarely unvarnished and, far from post-authentication, visitors detect restoration and the myriad traces of an unbroken line of past craftspeople who co-create their experience of authentication. As Chapters 10 and 11 also indicate, there is a close relationship between tourist, craftsman and structure, which reflects how ideas, people and places are connected in a more-than-human, networked world. Finally, it is important to note that the example of the 'Other Venices' fully demonstrates how places replicate other places. As Eco (1986, p. 19) indicated, the implication is that these copies may remove the desire for and finally supplant the originals that they emulate, transforming reality into hyperreality.

4 Cities as stage sets

From gentrification and festival market places to hipster places

Introduction

This chapter identifies different fusions of fake and real in a variety of gentrified environments, illustrating how contemporary urban 'real-fake' (Fjellman, 1992; Hannigan, 1998) places practise staged authenticity. Cities are increasingly intended to influence and reflect the taste of discriminating tourists, who read the semiotics of gentrification like texts (De Certeau, 1984; Donald, 1992) – or, perhaps, increasingly, style bibles – as they experience and authenticate the *mis-en-scène*. Richards and Palmer (2010) argued that cities resemble stage sets, while Lovell (2013) identified many idioms of urban theatricality, including postmodern architectural design references, which act as backdrops for street furniture, props and actors. Examples of staged 'front regions' (Goffman, 1959) include 'festival market places' (Bryman, 2004; Ritzer and Liska, 1997) and 'neo-bohemias' (Lloyd, 2010), both of which provide aesthetic façades for 'heritage shopping' experiences (Bryman, 2004, p. 35; Orbaşli 2000; Crawford, 1992, p. 16).

The chapter shifts the discussion towards 'hipsterised places', where industrial heritage traits and cultures, including symbols of the working class, slow anachronistic crafts processes and vintage products, have been appropriated and repurposed. The hipsterisation of places such as SoHo and Williamsburg in New York, Wicker Park in Chicago and Shoreditch in London (Lloyd, 2010; York, 2016; Zukin, 2010a) is highly apparent, to the extent that Shoreditch has been described as a 'hipster theme park' by York (2016), suggesting a new urban theatre for tourism. In these examples, the front stage has been refashioned to resemble a 'backstage' area (MacCannell, 1973) by reinjecting signs of industrial heritage. The discussion builds further on previous chapters, suggesting that the deeper origin of gentrified places is sought by tourists who 'mechanistically authenticate' (Lovell, 2013) the symbols of blue-collar workplaces and artisanal products and practices which they encounter. The chapter positions staged authenticity within Lovell's (ibid.) process of 'real to fake to real', exploring how vestiges of the life-story of gentrified places are grafted onto the present: urban memoirs traced over present scenes by historical prompts such as the Stonewall Sign (see Figure 4.1).

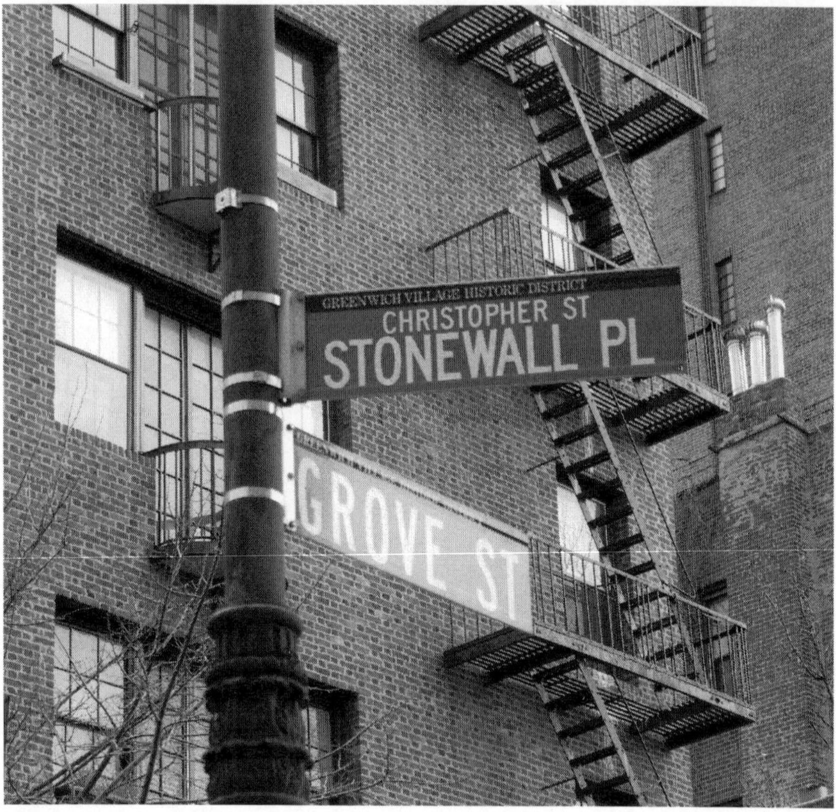

Figure 4.1 Character through detail: New York fire escapes

Source: Tim Bolwell

The gentrifying gaze

This chapter explores the 'gentrifying gaze', which fits into the category of 'cooler authentication' (Selwyn, 1996; Cohen and Cohen, 2012) by valuing original and repurposed heritage in a heady combination of commercial nostalgia. Both Zukin (1996) and Boyer (1996) suggest that the gentrification of districts in New York provides evidence of the quest for nostalgia-based authenticity among an alienated, growing middle class. It has been argued that the middle class and creative class use goods as symbols of happiness and cultural capital to confer status and distinction (Bourdieu, 1984). Baudrillard (1981) proposed that the search for distinction is the result of a 'lack', which echoes MacCannell's (1973) theory of a cyclical tourist quest for authenticity. This is expressed by Fitch (1990, p. 325) as 'the need for alienated peoples to re-establish some experiential contact with the material evidence of their own past'. York (2016) suggests that this lack is due to the growth of the service and knowledge economies, which distance people from manual labour and crafts, indicating that tourists are (desperately) seeking to sample a working-class, authentic past.

Characters in novels tend to have 'back stories' (Vogler, 1999, p. 94) and so do places. Discovering what McCarthy (2008, p. 536) has described as the 'micro-histories' of heritage places provides tourists with another backstage glimpse. Stories (often repurposed) are attached to places: from the transformational memoirs of neighbourhoods such as the Meat Packing District in New York to diaristic details, for example, the brown paper bags still used by a family iron-mongery business. Gentrification repurposes post-industrial districts, embellishing buildings, food and drink with value-laden 'spatial narratives' (Meethan, 1996) with shallower roots, replacing older chronicles of place.

As debated in the previous chapter, insights into craftsmanship are of interest to tourists as a way of mechanistically authenticating places. Everett (2012) debates the placemaking attributes of food tourism, emphasising a growing interest in production and methods of cooking. From the farm shop to the farmer's market to the vegetable box or allotment, the fascination with local or home-grown food is at its peak. The hipsterisation of some places provides visible signs of staged authenticity in the craftsmanship involved in artisan food and drink, percolating through into what York (2016) describes as 'micro-connoisseurship' practised by tourists and local people. Cool authentication is warmed and slow-dripped; cold-brew coffee is a front-stage performance, as are displays of micro-connoisseurship, but, as the chapter will explore, slow-dripping cool authenticity may actually warm it.

The middle classes have high expectations of places; their mobility and a 'certain level of shared cultural experience and memory' (Benton and Cecil, 2010, p. 25) can cultivate a higher state of aesthetic awareness (Maitland and Smith, 2009) and a discerning eye, meaning that they are more critical of their environment. Discussing New York, in *Naked City: The Death and Life of Authentic Urban Places*, Zukin (2010a, p. 20) comments that 'we can see "authentic" spaces only from outside them. Mobility gives tourists the distance to view a neighbourhood as connoisseurs… especially when we look at a rundown neighbourhood we ask, Is it interesting? Is it gritty? Is it "real"?' As Lovell's (2013) research indicated, people are interested in the origin of places; gentrification involves reinjecting authenticity and cultivating nostalgia through 're-heritagization' (Ashworth, 2008, p. 261). The urban studies expert Jane Jacobs (1961) detailed how placemakers attempt to locate the local within the global and find the village within the urban. Tellingly, Zukin (2010a) observes that neither Robert Moses nor Jane Jacobs used the word 'authenticity' when writing about New York design.

Emergent authenticity: real to fake to real

This chapter explores the effect of urban renewal and revitalisation on authenticity. Lovell (2013) defined a process of 'real to fake to real' in historic cities as they recalibrated their sense of place while experiencing sudden accelerated gentrification after decades of design conservatism. Zukin (2010a, p. xi) depicts the tension between 'origins and new beginnings' in post-industrial landscapes: 'the features that every generation views as "original" because they have been there throughout their lifetimes and features that each new generation creates on their own'. As demonstrated in Chapter 2, tourists seek myths of origin and it is important to retain the layering of narratives.

The phenomenon of 'emergent authenticity' (Cohen, 1988b; Coleman and Crang, 2002) is evident in environments ranging from seaside resorts to village hamlets. Bruner (1994, p. 407) stated how 'all cultures continually invent and reinvent themselves', implying that places are in a constant state of flux and adaptation. Depreciation can work in reverse, reforming opinions, the effect of gentrification is the revalorization of everyday items such as collectables, kitsch, or vintage. Emergent authenticity was particularly demonstrated in the growth of industrial heritage tourism during the twentieth century. Lynch (1972, p. 43) describes how the Ironworks at Colebrookdale moved from being 'backward' to being considered 'historic', demonstrating the malleability of social taste and memory. Industrial heritage tourism has been though incarnations of popularity; for example, 1993 was the Year of Industrial Heritage (Goodall, 1993; Jansen-Verbeke, 1999; Timothy, 2001) and during the Millennium redevelopments emblematic industrial sites such as the Tate Modern in London and the Baltic in Newcastle were adapted as art centres. The mainstream popularity of industrial heritage symbols and processes as part of urban gentrification indicates a more diffuse life-cycle upsurge.

Cities are not immune to changing trends, moving in or out of fashion because of factors such as mass-media exposure, as described by Bayley (2007), 'Fashion made Bath famous, but, as Coco Chanel knew, fashion is what goes out of fashion.' The totemic Sheffield Park Hill Estate epitomised emergent authenticity; opening as a new vertical city, it went through a cycle from success to deprivation, was abandoned and fell into disrepair. English Heritage subsequently reappraised the concrete structures, which were converted once more into flats. According to Historic England, 'Park Hill in Sheffield is our current poster-boy Grade II* listed regeneration project… [it is a] product that is contemporary in design terms – and thus palatable to the requirements of the modern marketplace – but that avoids pastiche' (Lynch, 2014).

Cities have always evolved, from the slum clearance of Haussman's boulevards and erection of monuments in Paris, to the laying out of the City Beautiful grid model of Washington. The urban environment can be viewed as more dynamic, with a faster rhythm and a bricolage of influences, and thus broader brushstrokes are used when planning change. Multiple factors have significantly contributed to the creation of the contemporary urban environment, including globalisation and deregulation. The building of the superhighways in America in the 1950s created a new consumption ethos in direct competition with the shopping core of cities. The creation of out of town shopping centres, supermarkets and hyper-markets led to the 'doughnut effect' apparent in what Pallagst *et al.* (2013)

describe as 'shrinking cities', where people move from the city centre into the suburbs or new high-rise builds, a movement attributed to population loss. Lloyd (2010, p. 55) discusses how bohemias continue to draw on the romance of the artistic ghettos of 19th Century and 1920's Paris, which were crucibles of creativity and generated what could be described as the source aura of bohemianism. The spirit of Montmartre was reprised by the Beat generation in the 1950s in Greenwich Village, which established a more recognisable version of the urban hipster. The authentic self was expressed through the intrinsic benefits of creative freedom and, as discussed in the introduction to this book, linked to the existentialism of Sartre. Ley (1996, p. 199) describes how artists arrive first in the abandoned inner cities, pioneers starting the process of gentrification along with those who need cheap accommodation and business premises, creating bohemias.

Cohen (2007) suggests that sometimes the framing and reframing of places can be inauthentic, but Hull's post-industrial decline and subsequent designation as Capital of Culture 2017 has given it a new culture to grow into, while still valuing past achievements. The aura of industrial maritime heritage has been revived by destinations such as Portsmouth and is highly evident in numerous museums and visitor attractions; but for many post-industrial places, cultural representations of industry are simply fading echoes of the past. The new meaning and economics of post-industrial cities was addressed by Florida's influential 2002 text, *The Rise of the Creative Class*, which advocates designing places for the knowledge economy and creative class. Lloyd (2010, p. 70) describes Florida's realisation that bohemians no longer saw themselves as alienated or marginalised, but that their creativity was now mainstream and compatible with conformity. Muggleton and Weinzierl (2003) suggest a post-subcultural approach, indicating that subcultures are now motivated by style and a desire to be individual rather than rebellious; causing the mainstreaming of design for the creative class, commodifying forms of self-expression as hipsterisation and ultimately designing out bohemia and designing in slower, simulated ethnic neighbourhoods (see Figure 4.2).

Lloyd (2010, p. 257) argues that:

> the creative cities discourse in planning has also developed close ties to the design movement of New Urbanism, which emphasises density, pedestrian culture, human scale buildings and mixed primary uses, the sort of landscape Florida believes best nurtures creative impulses among city residents.

Forging authentic links between creative cities and their community involves commoditising the aesthetics of sense of place and local cultural capital. Showcase service amenities are designed to attract the knowledge economy and cater to the creative class and cultural consumers (Florida, 2002; Maitland, 2007; Zukin, 2010a; Salazar, 2010). This holistic form of inner-city gentrification was realised in its modern form in Portland Oregon in the 1990s, home of the first postmodern architecture, the Portland Municipal Services building built by

Michael Graves. The key features of regeneration in Portland involved the reuse of the inner city, reconnection of the tram system, a signature attraction, Border's bookshop – a marker of the proto-knowledge economy – public art, adaptive use of warehouse space, a waterfront development and a relationship to the nearby Seattle Grunge scene, all of which gave a feel of old-fashioned, politely romantic, hip modernity. It was perfectly constructed to attract young professionals from elsewhere in the USA.

Sanneh (2016) detailed how 'the word "ghetto" has come to sound like an indictment of a people as well as of a place' and, in the same way, the word gentrification is synonymous with ideas of cultural appropriation and neighbourhoods upgraded for the bourgeoisie. Gentrification is value-laden, it implies that authenticity is eradicated in real terms, then commodified as local people and businesses are priced out of the market. In some instances, this process results in what Lloyd (2010) calls a 'neo-bohemia', where the actual elements of bohemia, poverty, dereliction and artists have been eliminated, but the counter-cultural reputation of the original place is retained. Neo-bohemias emerged in a process Zukin (2010a, p.51) terms 'from gritty to cool', which is epitomized by the hipster district of Williamsburg:

> Nouveau grit not only describes Williamsburg's revival; it also applies to the rebirth of San Francisco south of Market Street during the dot com boom and the Seattle of Starbucks and grunge, as well as to the revival of a small number of industrial neighbourhoods… Gritty's appeal was in the postindustrial spirit of the times and in the symbolic economy's ability to synthesise dirt and danger into new cultural commodities.

As described earlier, tourists and local people alike seek the backstories of places as forms of authentication – for example, Lloyd (2010) describes 'grit as glamour' to denote its allure. The embodiment of grit can be traced to the smells of the city, as McLean's (2016) smellwalks and smellscapes delineate, from Amsterdam to New York, from marijuana to dampness. As lifestyles change, it could be argued that the smells of working markets and industrial heritage, which are perhaps not as attractive to the modern 'sensory' palate, could be lost, changing the embodied view of places. Relph (1976 [2008], p. 71) described how the sensation that places are lived-in and subverted by hosts can make them seem authentic to tourists, offering a modern backstage glimpse of local people as living graffiti, which, ironically, gentrification processes then obliterate. The life-story of places is told through eclectic styles, but when 'real' elements of human graffiti are erased, they are replaced by designed edginess, in the form of curated art murals and graffiti in hipster places such as Shoreditch (see Figure 4.3).

Huxtable (1997, p. 27), referring to the conservation of heritage, stated: 'What are gone are the cumulative clues, the patina of age and use, the sense of "others" – that essential, irreplaceable quality that Stephen Greenblatt has so insightfully called "resonance".' Huxtable goes on to discuss inauthentic 'hollow history', a phrase which could be applied to the structured absence of original local people;

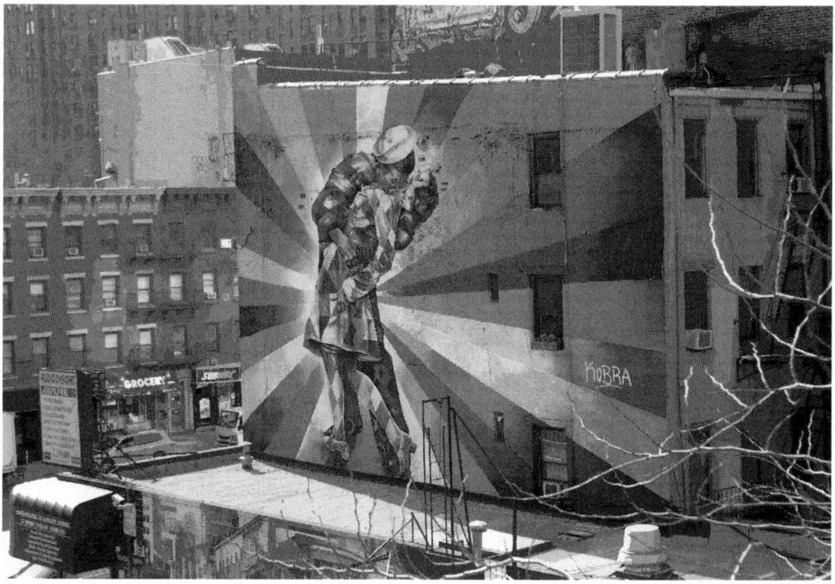

Figure 4.2 Kobra mural, New York

Source: Tim Bolwell

the place of origin, or backstage, retreats further with those who have left, leaving hipster tourism in its stead. The paradox is that while the edginess and 'rawness' attracts the pioneer artists, they are edged out as places are reconfigured for the creative classes (Ward, 2016; Mathews, 2010), a phenomenon which is discussed further in the chapter on art.

Attempts to infuse places with bohemian associations by drawing on the aura, uniqueness and singularity of heritage are not always successful. Lovell (2013, p. 274) describes the case of Palace Street in Canterbury, the site of independent shops in historic buildings and which aspired to cultivate the uniqueness of The Laines in Brighton. The Palace Street traders created a 'King's Mile' identity, created by local traders to draw visitors off the High Street. The name refers to the mile that Henry II walked in penance for the murder of Thomas Becket. There is no evidence that the king walked that route and it is not a mile. The city centre manager in Canterbury also implemented a monthly market called Bohemia in Palace Street, using cultural agency to cultivate the impression of a neo-bohemian area. While bohemian was an adjective also used by placemakers in Lovell's research, tourists described Palace Street as 'quaint' (ibid.), indicating that place evokes its own atmosphere and narrative.

Gentrification ripples outward from urban centres, as the creative class, metropolitan elite or yuppies move in (Lloyd, 2010) and the search intensifies for new backstage zones. As Maitland and Newman (2014) detail, areas off the beaten

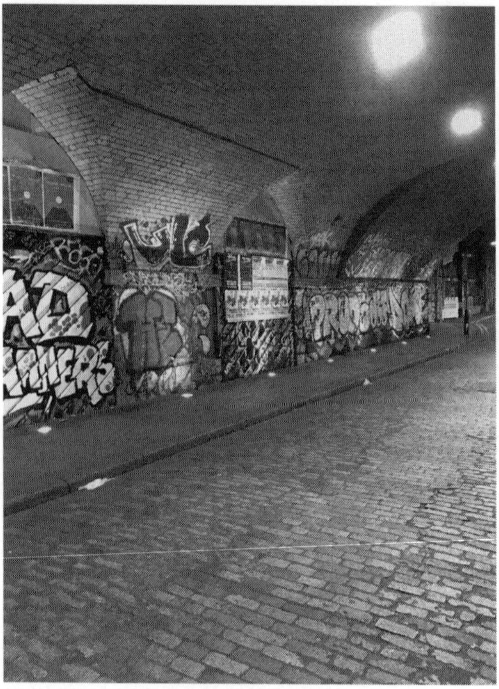

Figure 4.3 Shoreditch graffiti, London
Source: Andrew Robertson

track of landmark tourism, such as Islington in London, become increasingly popular as tourist attractions. Acts as simple as office work in these neighbourhoods can seem performed, yet perversely make places seem more like backstage environments for their very ordinariness. Baudrillard (1981, p. 104) describes the performance of the ordinary: 'The freeways, the Safeways, the skylines, speed, and deserts – these are America, not the galleries, churches and culture.' Authenticity is the everydayness.

Hipstorical places

The authenticity soup involved in the infinite combinations of old, new, reproduction and recreation provides an infinite mix of modern interiors in heritage buildings, or heritage interiors in buildings with modern façades. Adding a dash of original, crafted, traditional products further spices the debate about staged authenticity in the urban environment. It could be argued that heritage shopping includes repurposing, appropriating, reclaiming and reinventing existing buildings until, in the same way as the Ship of Theseus in the previous chapter, it is difficult to know which aspects, if any, are truly 'original'.

People authenticate places in a variety of ways, including emotionally, bodily, performatively and mimetically (Bagnall, 1996; MacDonald, 2009; Pink, 2007; Smith, 2006; Zukin, 1996). Another aspect of 'mechanistic authenticity' is the consumption of the places they visit through their stories. Many gentrified areas have repurposed and revived de-industrialised districts, buildings, food and drink, giving them value-laden backstories and rehabilitating the characters and story arcs of places. Different actors and networks layer their self-histories, providing opportunities for micro-connoisseurship (York, 2016) in areas such as Spitalfields (Maitland, 2010) or Borough Market. While the industrial heritage era remains in the past, artisan products enable the performance of mechanistic authenticity at a granular, detailed level, allowing tourists to express their discernment, style and taste. All of this contributes to what Jansson (2010) describes as the sediment and Boho-luxe and accumulation of authenticity in cities.

Modern cities were designed to be consumed by the creative classes (although not poor artists); competing with the Haussmanian boulevards of capital cities such as Paris and Washington to invite the gaze of the global society. Places of imagination are peopled by characters and archetypes, one of the most flamboyant of which is the flâneur or flâneuse, traditionally defined as the poet, spectator, romantic figure, or vagabond – a motif in Baudelaire's poetry (Benjamin, 2002, pp. 417–20). White (2008, p. 45) links flâneurie to the taste-consciousness of Bourdieu. It is important to include the flâneur in any discussion of authenticity and place because, while this urban character seems arch and ironic, their emergence was linked to the loss of the rural idyll. The flâneur was a reaction to the nineteenth-century urbanism of Paris: 'a new Romantic conception of landscape emerges – of landscape that seems, rather, to be a cityscape, if it is true that the city is the properly sacred ground of flâneurie' (Benjamin, 2002, p. 420). The figure of the flâneur demonstrates how the romantic gaze can be applied to man-made, urban spaces and their *mis-en-scène*. The setting for flâneurie was what Benjamin describes as the 'colportage phenomenon of space' (the word 'colporteur' means peddler). He suggested that the flâneur (or flâneuse) is at home in 'multiplicity', or chaotic, 'seething' situations. Flâneurie uses Paris as a stage set, attributing imaginary lives to passers-by; wrought-iron arcades such as the Passage des Panoramas becoming 'a landscape made of living people' (ibid., p. 240).

The object of the flâneur is differentiated from that of the tourist (White, 2008, p. 25). White (ibid., p. 47) states that both flâneurs and tourists are searching for authentic, obscure or evocative references and oddities, but where tourism looks for more obvious landmarks, flâneurie seeks romantic, eclectic or surreal atmospheres; in other words, unstaged backstages. Authentic ways of consuming cities on foot include the 'discursive walking' of the flâneur, which resembles the 'conceptual' critical walking methods of situationists and psychogeographers (Edensor, 2010, p. 74). Coverley (2006) details how psychogeographers treat walking 'as acts of subversion', choosing to wander in 'marginal and forgotten areas', using walking as placemaking. In addition, the public realm in new, themed consumption spaces such as neo-bohemias and festival market places is tailor-made for sauntering. Adoption of the role of flâneur by tourists is encouraged by the number of exterior/interior effects created by these spaces and their cinematic spectacle, semiotics and aesthetics. Benjamin, (2002, p. 419) describes flâneurs as euphoric and 'intoxicated' by their urban surroundings, which suggests 'hot authentication' (Cohen and Cohen, 2012); flâneurs pursue the psychoheritage trail of intangible clues about the past.

Hipsterised, gentrified places attract the flâneur; they are deemed more culturally authentic and therefore more valuable if they were once sites of artistic poverty or manual industry, such as the Meat Packing District in New York. The place of origin, the slaughterhouses, are still somehow implied, or made explicit through residual, visual cues in a meta-historical effect, the appropriation of working-class values authenticating the hipster tourist gaze. The typologies of hipster, tourist, flâneur (and flâneuse), creative class and metropolitan elite increasingly overlap. They all authenticate places with their gaze and creative places are increasingly designed to authenticate *them* (and their sense of distinction). Roodhouse (2006, p. 22) identifies 'cultural iconographic regeneration' as including single flagship projects such as the Whitney in New York and the Turner Contemporary in Margate. The Whitney Museum of American Art was constructed by Renzo Piano in 2015, with spectacular views of the city, located at the start of the High Line, the converted elevated railway in New York, which has been replanted as a garden, art venue and city viewing platform. The museum is situated beside the High Line (the rail track converted into a park) with its panoramic views, public art and pop-up cafés and is also a short walk from the Meat Packing District, where the adapted warehouse building of Chelsea Market contains Boho-luxe boutiques, all of which feed into and reinforce aspects of the hipster tourist gaze.

Detail and micro-histories are important to hipster tourists seeking to authenticate places. Independent retail is a feature of gentrified places and is perceived as adding social value, sense of place and prestige. For example, Zukin discusses how family-run businesses are part of the authentic mix in neighbourhoods (2010a) and Lovell (2013, p. 253) recounts how Birches ironmongers in Shrewsbury provides old-fashioned, performed service touches such as goods sold in brown paper bags. Littrell *et al.* (1993) examined tourist perceptions of the authenticity of craft souvenirs and the research suggested that tourists denoted objects authentic if they met criteria such as a connection to past traditions. The use of old-fashioned, brown paper bag is perceived as adding value and ambiance rather than being as anachronistic, because the family has perpetuated this custom. Lovell (2013, p. 253) describes how the presence of second-hand and independent bookshops authenticate some places for tourists – for example, signature attractions such as King's College Books in Cambridge and Blackwell's in Oxford. Bookshops are integral to these university cities and buying the book is an act of authentication, encompassing the purchase of a piece of historic

Oxbridge intellect, much like the City Lights Bookshop in San Francisco, with its strong aura of the 'Beat' authors.

Food, drink and shopping and the artisanal production of luxury handcrafting of goods is at the centre of gentrification and is unpicked through the 'mechanistic authentication' (Lovell, 2013) of places by tourists who wish to understand more about their provenance and story. The 'historaunt' (Josiam *et al.*, 2004) historical café or pub are common tropes, the adaptation of heritage buildings as modern eateries, fusing different timescales. There is a difference between hipster cafés and more traditionally established historaunts such as Betty's tea shop in York, which provide a further way of consuming space through 'signature place' and can be said to have a singularity or originality about them. The location of Betty's in a historic city, on a historic crossroads of a Roman road, is authentic, and the building and interior are art deco. Betty's fuses the Swiss origins of its bakery with Yorkeshireness: the produce is local, loose leaf tea is served in a pot and cakes are locally sourced (Wild, 2007). Stobe's Coffee House in Lincoln occupies a timbered, black and white Tudor bridge over the river. This is another example of a signature historaunt, due to the blended elements of location, traditional activity, historic building and the smell of fresh coffee adding a multisensual aspect to the gaze. The restored Princess Louise gin palace in Holborn exemplifies the concept of the iconic 'histopub' (adapted from Josiam *et al.*, 2004), recreating its original Victorian character 'snob screens' of etched glass, sheltering the private socialising of the middle class from the working class in the bar. These examples are significant because they have a uniqueness to them that draws tourists and local people and cannot easily be replicated in hipsteraunts. Betty's and the Princess Louise both cherry-pick the best of both the past and contemporary eras in a 'time collage' (Schlereth, cited in Lippard, 1997, p. 93) based on continuity. While the provenance of establishments such as Betty's is evident, gentrified cafés and bars tend to promote the local production of their food and the craft processes of their microbreweries as the (slowish) source of their authenticity. Traditional artisanship sometimes agglomerates as a theme: Brighton has a 'chocolate cluster' of shops, including Hotel Chocolat, Montezumas and Choccywocky DooDah. Traditional, mimetic acts of heritage craftsmanship, such as chocolatèrie, provide resources for mechanistic authentication. Some shop and café interiors indicate their sense of heritage by resembling workplaces, with exposed brickwork, distressed walls and industrial elements such as lightbulbs on display. Lovell (2010) argues that gentrification stylisation seeks to emulate the atmosphere of the 'originals' – for example, the destination alleyways and lanes such as North Laine in Brighton, St Edward's Passage in Cambridge, The Shambles in York and the Ginnells in Lancaster. Bashful Alley in Lancaster is one of the Ginnells off the beaten track, which dates back hundreds of years and contains shops, a café and a pottery.

It is important to note that the mechanistic authenticity of some places is paradoxical to the point of parody. As discussed earlier, Portland Oregon was an early example of gentrification and is a prototype of fashion. Both the environment and the reputation for grunge (Zukin, 2010a) made it attractive to young professionals as a place to live and work. It has been portrayed in the series *Portlandia* as a nexus for hipsters, a reputation which was first planted in its initial gentrification. Sketches such as 'Is It Local?', where the hipster couple examining a restaurant menu go to visit the chicken which they are about to eat on the farm, show mechanistic authentication at its most exaggerated.

The hipsterisation of places is characterised by the Millennial generation's joyous rediscovery of the crafts production of previous generations and their celebration of anachronism. This mirrors the periodic folk movements which resist modernity, from William Morris's predication for the rediscovery of crafts, to the Art Nouveau focus on the organic in architecture to the hippy fashions of the 1970s, all of which were commodified as style statements and are considered further in Chapter 7. As Relph (1976 [2008]) discussed, nostalgically designed places tend to seem inauthentic and there is a Heath Robinson 'retrochic' (Samuel, 1994, p. 83) quaintness to hipster places that, while the crafts are authentic, makes their practice arcane, which for hipsters is the attraction. Retrochicery is embraced by iconic hipster shops such as Barn the Spoon in Hackney, where the owner whittles (and teaches the whittling of) hand-carved utensils using timber from Tower Hamlets Cemetery Park (Galliver, 2013).

It is no coincidence that hipsterism has developed during a time of increased technological dependency. The artisanal production of food and drink, popularity of allotments, sit up and beg bicycles, cycling (and cycle polo), the wearing of working-class clothes and explorer beards all echo the past and are attempts to infuse lifestyles with meaning beyond the digital. Benjamin (1936 [2008], p. 9) argues that there is a 'decay' of aura through reproduction: 'the desire of contemporary masses to bring things "closer" spatially and humanly, which is just as ardent as their bent toward overcoming the uniqueness'. Fatigue with methods of mass-reproduction could be said to rest at the heart of the handmade movement, but, ironically, the first impulse of many people regarding a craft item which they have made is to digitally photograph it; a seeming inconsistency which is examined in Chapter 9.

Hipstorical public realm

Using mimetic design references in the public realm could be argued to constitute staged authenticity. For example, mimetic, faux Victoriana street furniture lends the streetscape the sense of 'buying second-hand time' (Januszczak, 1996, in Hewison, 1987, p.134). Hewison (1987, p. 133) contends that 'recycled imagery' actually creates a distance between ourselves and the past. Paying attention to detail adds to the ambiance of places – for example, the historic lampposts on Lendal Bridge in York which offset its 'big picture grandeur' (Lovell, 2013). Street furniture may underpin the cultural narrative; some of Winchester's bollards have been decorated in a cod-Mondrian style to prompt people that Winchester is a 'Creative City' (Landry, 2000). This contrasts with Chester's medieval-style bollards, which reflect the timbered buildings, the bollards adding a further prop to the stage set.

Tourists perceive theatricality and staged authenticity in the props, such as swinging signs and directional fingerposts (Lovell 2013). They add to the mis-en-scène, as do the proscenium-like, giant, red-coloured Tsuzumi gate outside Kanazawa Station in Japan, based on traditional hand drums and the gateways to China Town in San Francisco, Newcastle and London. Gentrification uses objects to 'anchor' time (Tuan, 1977) and placemaking caters to the anachronistic desires of hipster tourists by providing nostalgic, stereotypical scenery. Lovell (2013, p. 268) discussed how tourists photographed the archetypal emblems of culture in historic cities – including pubs, beamed houses, buses, phone boxes, post boxes and fish and chips – their gaze warmly authenticating what they described as the 'real Britain' or 'real England' in a playful, ironic, way (Larsen and Urry, 2011) recording the recent alongside the ancient past, the 'costume' of the policeman alongside the monk. Preserving the symbols of obsolescent modernist culture, such as telephone boxes, therefore implies that when function ends, staging begins.

Pastiche

Any discussion of authenticity and place should include pastiche architecture because it plays such a central role in our everyday environments. Authenticity has a relationship to time, as discussed in the previous two chapters, and confusing the timescales of places can rob them of their overall sense of originality. Design tropes and combinatory forms, such as the brick-clad façades on new buildings, extend the number of eras and styles which could be considered 'heritage' (Richards, 1996, p. 273), making the actual historic environment seem less perishable and rare. Bruner (1994, p. 401) described qualities of 'verisimilitude' and 'genuineness', but pastiche epitomises postmodern authenticity. Urry (1995, p. 219) contends there is 'an increased significance of pastiche as the past is sought through images and stereotypes which render the "real" past unobtainable and replace narrative by spectacle'.

Pastiche has its critics. Jameson (1998, p. 5) describes pastiche as 'blank parody, parody which has lost its sense of humour'; it could be said to dilute authenticity. Pastiche is also associated with design conservatism in historic environments. Christopher Frayling, Chairman of the Arts Council, commented on the rejection of the Dyson design school by Bath City Council in 2007:

> The situation is very sad. The WilkinsonEyre Dyson building was a very distinguished piece of architecture and the Holburne extension as well. Meanwhile the pastiche Georgian SouthGate shopping centre is going up. It seems that if it's custard colour, it's good. I do think the preservation people wield undue influence.
>
> Bayley, 2007

Rather than creating a sense of placelessness, it could be argued that the use of pastiche creates a sense of timelessness, threatening to mire places in what Jameson (1985, p. 6) termed the 'perpetual present'. However, pastiche was always aspirational, as manifested in the Tudorbethan style, and can add to overall atmosphere; for example, the City of London, benefits from the juxtaposition between old and new.

Pastiche makes it hard for tourists to read and therefore authenticate places. Hannigan (1998, p. 68) described the style of 'authentic reproduction' in the festival markets of the 1970s and 1980s, evoking nostalgia through old-fashioned detail and 'gingerbread architecture'. Pastiche uses a 'distoric' approach; Fjellman (1992, p. 60) referred to 'Disney realism', where the past is commercialised for the present. There are many forms of McDisney urban style (Bryman, 2004; Sorkin, 1992; Ritzer and Liska, 1997). For example, Longmarket Square in Canterbury, constructed in 1995, uses vernacular details of red-tiled roofs and mullioned windows on a larger scale than the pilgrims' inns which are characteristic of the heritage city. When it opened, the reactions were mixed, as described by a placemaker in Lovell (2013 p. 282):

> Longmarket was designed on a fairly traditional basis with a sort of medievally type character and I always remember when the hoardings came down the public said 'what a great looking scheme' and all the architects said 'it looks Disneyesque'.

Tension between conservation custodianship and Disney authenticity mean that historical references in design could appear artificial, but as a Canterbury planner observes (Lovell, 2013, p. 283):

> you have gothic revivalism, you have Georgian architecture, so most of Georgian Canterbury is just re-fronted medieval buildings because it was a fashion at the time so if you look behind most Georgian façades you will find a medieval building, so was that Disneyesque?

In other words, gentrification is nothing new; places have always been imitative.

Festival market places

Another important permutation of fake and real in the urban environment are areas known as 'festival market places' (Bryman, 2004; Ritzer and Liska, 1997). These are 'real-fake' exemplars of staged authenticity, designed by private developers for a mass audience, but striving to avoid the 'clone town' (Conisbee *et al.*, 2004) placelessness (Relph, 1976 [2008]) of modern shopping centres by repurposing industrial heritage sites such as Covent Garden in London. It is important to note that heritage has been argued to act as a façade for shopping (Bryman, 2004, p. 35; Orbaşli, 2000; Crawford, 1992, p. 16). Bryman (2006, p. 35) argues that festival market places are 'areas with "real" heritage value turned into shopping locations that feed off a sense of a valued past and thus function essentially as themed malls'. To replace the aura of originality, festival market places use reproduction and amplification, turning up the volume on urban design, using an illusory learning-fun atmospheric mix, which Urry (2002, p. 136) has described as 'edutainment'. Benjamin (1936 [2008], p. 9) discusses the effect of reproduction on aura: 'the desire of contemporary masses to bring things "closer" spatially and humanly, which is just as ardent as their bent toward overcoming the uniqueness'. The amount of replication in the built environment can evoke a sense of placelessness (Relph, 1976 [2008]), or a 'global sense of place' (Massey, 2005, p. 192).

The gentrifying gaze recognises the comforts of modernisation, attributing value to residual signs of authenticity. Research suggests that shopping in heritage buildings seems to increase the value of the items purchased (Lovell, 2003), as if the goods have taken on the value of the place in 'adjacent attraction' (Crawford, 1992, p. 14). Some festival market places use adjacent attraction on a wider placemaking scale to tell the perennial story of industrialisation; the South Street Seaport area of New York has been portrayed by Boyer (1996, p. 182) as 'an imaginary historical museum... surrounding the spectator with an artfully composed historic ambiance'. Bryman (2004) examines falsified environments in modern shopping centres, for example the New Orleans Quarter in Manchester's Trafford Centre and modern facilities that amplify the historic design of the buildings where they are situated, such as Faneuil Hall in Boston.

Festival market places are performance stages; the type of experience they provide takes them several steps away from originality, offering a 'flâneur approach to the busy metropolitan life' (Ebner, 2008, p. 187). Boyer (1992, p. 184, in Sorkin, 1992) sets out how cities are designed to create theatre and tableaux:

> these newer sites are laden with historical allusions to the traditional vision of the city: a coherent place of intimate streets, lined with small-scale façades and shopping arcades, ornamented with signs, punctuated by open spaces, trees, lampposts and benches. The aim is theatrical: to represent certain imaginary prosceniums in order to conjure up emotionally satisfying images of bygone times. Architecture and the theatre use similar means to design places of pleasure and spectacle, manipulating scenery, ornament and façades, to underscore the sentiment of their play.

Examining the history of festival market places further, the 'circulation' of global designs, brands and ideas was influenced by private-sector ownership of land in the centre of cities. Companies such as the Rowse Corporation employed a notoriously formulaic approach to their waterfront redevelopments in America. New shopping centres built after the Millennium and before the economic recession of 2008 provided an expected mix of brands contained in purpose-built space, with chain stores in shopping centres adding an element of the 'brandscape'

(Klingmann, 2007). Festival market places are designed to incorporate the new public squares, an idea recycled from nineteenth-century ideas of spectacle (Boyer, 1996, p. 187). 'Like civic monuments and memorials, commercial environments today constitute loci where people connect with core cultural ideas' (Klingmann, 2007, p. 223).

Festival market places typify the control expressed in Ritzer and Liska's (1997, p. 97) theory of 'McDonaldization', which they define as a 'modern grand narrative' based on rationalism. However, even the most strictly choreographed and controlled themed environments become the sites of authentic experiences and acts of resistance. This type of environment is also described by Richards and Wilson (2007, p. 14) as 'efficient, calculable, predicable' and 'dominated by technology' and 'serial reproduction'. McDonaldisation is a way of structuring reality, privately owned shopping spaces have their own rules and security, with the intention of providing a risk-free environment and the 'soft control' of guards, guides, CCTV and specific rules or diktats creating a strong sense of scrutiny, banning normal tourist activity such as photography (Edensor, 2001, p. 64; Ritzer and Liska, 1997, p. 106). A desire for urban connectedness and business investment means that some places try to be all things to all people and, in the process, dilute their sense of origin. Gottdiener (2001, p. 102) describes how some multifunctional quarters have become 'hybrid spaces,' in which a variety of acts of hyperconsumption can occur. Central to theming ideology is the notion of 'convenience', which is as embedded in city planning as the development of the evening economy and café culture, but Lynch (1972, p. 35) makes the point that the addition of modern conveniences can 'rupture the historical integrity'. As Ioannides (2009, p. 7) argues about standardisation,

> the situation is made worse by the fact that, with so little variation between the attractions many visitors begin to become bored with what is on offer; if you have seen a festival market place in one downtown environment perhaps you are not so eager to visit a similar setting in another city.

As Krolikowski and Brown (2010, p. 138) state: 'in the quest to become attractive to tourists, many urban spaces have lost what initially made them a focus of interest'. Festival market places may come to exemplify the effect of the diminishing returns of reproduction on the aura of places.

While these spaces are carefully staged and controlled, Richards and Wilson (2007, p. 15) observed that '[b]ecause urban areas are not seen as "authentic" landscapes in physical terms, they are forced to opt for experiential qualities, notably "liveliness" or "ambiance"'. There is a spirit to places which exists independent of placemakers, the 'sidewalk ballet' (Stevenson, 2003) and rhythm (Edensor, 2010) of the urban landscape. The rhythm of festival market places is speeded up by the implementation of atmospherics such as lighting strategies and bylaws like those designed to promote street cafés, all of which accelerate the spectacle. According to Bonn *et al.* (2007, p. 348) atmospherics use the 'Stimulus–Organism–Response' paradigm, employing:

> design factors (layout, color, clutter), social factors (persons within the store environment, such as employees and shoppers), and ambient factors (non-visual cues such as scent, sound and lighting), with the later addition of symbols and visual cues in the last category.

Augmented reality, chasing Pokemon and moving metro adverts all contribute to the sense of negotiable hyperreality. Post-industrial cities – for example, Manchester – use lighting, big-screens in live sites and other technical paraphernalia to enhance the urban space. The engineered environment strives to manipulate the visitor into engagement and, although this may be overt and contrived, the result may be a genuine emotional experience or 'gentrified feelies' (adapted from Fowler, 1989, p. 62, in Bagnall, 1996, p. 237). These aspects contribute to the real-fake atmosphere in the public environment.

Elements of staged authenticity such as atmospherics can affect the semiotic experience; the manipulation of the environment employed in 'fake-real' places may be pleasurable as we recognise that there are places designed for consumption. Lovell (2013) has described how staging in heritage cities can evoke perceptions of amplified, enhanced or decontextualised authenticity in tourists, which can anchor people in the present, causing a range of reactions, including irony. If visitors are aware of a sense of overt opportunism and artificiality, the real-fake experience can be a source of playfulness – for example, the Cereal Killer Cafés in Brick Lane and Camden, which draw on the nostalgia for childhood. Gentrification can provide opportunities for people to consume staged authenticity increasingly in their everyday lives as it ripples outwards into new neighbourhoods, but the question is whether the staging is so embedded that we expect our environments to entertain us.

It is important to note that gentrification has been argued to result in dissonance; it has been opposed by grassroots conservation movements in Boston and New York (Zukin, 1996; Boyer, 1996). In London, criticisms of gentrification solidified around the Cereal Killer Cafés (see Figure 4.4). The Brick Lane venue was the site of a protest against the oppression of poorer neighbourhoods by gentrification. Staged authenticity in gentrified areas is epitomised by the concept of 'grassroots globalisation' (Appadurai, 2000, cited in Bortolotto, 201, p. 98). Grassroots globalisation in areas like Wicker Park in Chicago, means that the artistic reputation is bolstered by staging, including pseudo-gritty bars and cafés, while on the street 'the homeless who may once have been part of the environment are made invisible' (Ioannides, 2009, p. 19). Gentrification is characterised by the creation of a utopic urban landscape where poverty is conspicuous by its absence – for example, the zero-tolerance policy in New York, arguably consolidated by Mayor Rudy Giuliani during his time in office, equating with Sibley's (1988) concept of 'purified spaces'. Once areas have been 'sanitised' and the unpalatable displaced, then new commercial distinctiveness can be reinjected; for example, Boyer (1992) detailed how, in New York, Disney opened a store in the South Street Seaport area that 'merchandised history'. Hannigan's (1998, p. 7) 'fantasy city', is the product of the new middle-class desire for risk-free experience, without poverty or

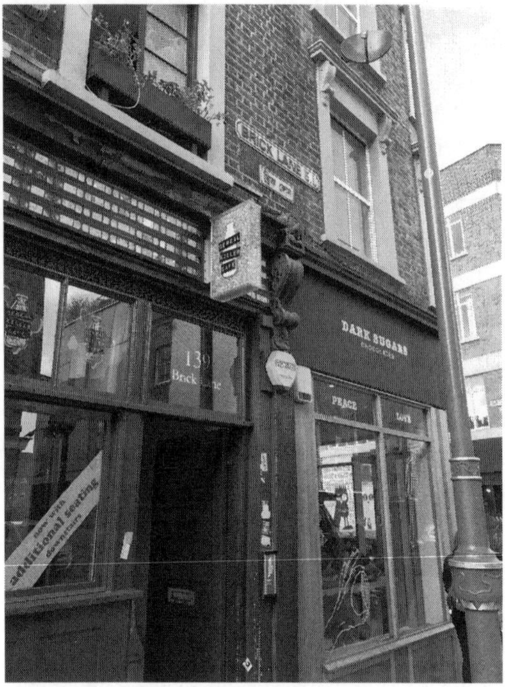

Figure 4.4 Brick Lane, London

Source: Andrew Robertson

evidence of dystopia. In a later work, Hannigan (2007, p. 54) observes that the commoditisation processes in neo-bohemias such as Wicker Park have created more similarities than differences between the 'riskless risk' of fantasy cities such as Las Vegas and the 'controlled edge' of the Creative City. These aspects do not pass tourists by; the search is on for new backstages and gritty bohemias.

Conclusion

The chapter has traced different forms of staged authenticity in the urban environment from gentrification and hipsterisation to festival market places. It has explored how gentrified places go through cycles of resistance, recalibration and emergence, from 'real to fake to real' (Lovell, 2010) or 'gritty to cool' (Zukin, 2010a) – appropriating, adapting and reviving original industrial heritage buildings, interiors, craft foods and cultural capital as symbols of place integrity. Gentrification forms a crucible, blending the auras of industrial heritage, blue-collar authenticity and bohemia recycled from nineteenth-century ideas of spectacle, superheated for global markets. Hipsterised places are more open to modern trends; they reflect a lighter, global form of authenticity, cultivated for flâneurie, the streetscene acting as an indicator of the health of the knowledge and creative economy. They are designed for a cooler (in both senses of the word) mechanistic authentication, but true coolness, like aura, however, evaporates in the visible effort to seek it, leaving only nostalgic sediment.

Criticisms of the authenticity of consumption spaces include their exclusivity, disconnection from grassroots movements, cultural imperialism, design banality and public art sponsored by 'corporate patrons' (Zukin, 1996, p. 23). The implication of gentrification for the aura of places is that if every place is made to seem more like its inner self by celebrating its local distinctiveness, this can eventually be self-defeating, leading to a proliferation of 'real cultural capital' (Richards and Wilson, 2007). The consequences of this are indicated by Coleman and Crang (2002, p. 3) as potentially causing 'a unity in diversity, or similarity at a higher level by making all these sites equivalent as potential places to visit'. Paradoxically, the more places strive to be different, the more they become the same.

Authenticity is not simply about place; it is closely aligned to our concept of time, which is anchored by the built environment. Pastiche architecture, historaunts, histopubs, and histo-shopping can create hipstorical places, resulting in heterotopias, or 'heterochrony' (Foucault and Miskowiec, 1986, pp. 22–7), suggesting that the enjoyment of mechanistic authentication by tourists may be blunted because they are unsure what is real. The weight of anachronism may eventually lead to gentrified areas becoming indeterminate places, losing the resonance, sense of origin and perennial backstories which attract the hipsters, flâneurs and creative class. Entropy may lead to a post-gentrified world, where today's pastiche becomes tomorrow's vintage heritage aesthetic and the artists return. To conclude, Soja (2010) uses the term 'Postmetropolis' to describe the palimpsestical layering of cities across different timeframes, which is apparent in the adaptive use of buildings. A pilgrims' inn in Canterbury was once used as a Swatch watch shop, bringing to mind Lynch's (1972) question 'What time is this place?' Perhaps the question should be rephrased, instead asking: 'What time isn't this place?'

5 Site-specific art

From magical realism and creative placemaking to simulacra

Introduction

This chapter examines forms of authentication which arise from the dialogue between art, artists, tourists and places. The chapter explores the 'soft' places of imagination and wonder, suggesting that art destabalises, reinvents, intervenes, immerses, articulates poetics, forges imaginaries and evokes the surreal. The Faraway Nearby was the title of the Whitstable Biennale in 2016 and referred to the phrase that the artist Georgia O'Keeffe used to sign off letters written from New Mexico to her family and friends in New York. The phrase also aptly describes how art can make familiar places nearby seem faraway, inducing strangeness and playing with the aura of locations. Art as defined by artists, commissioners and curators has the power to creatively authenticate places, telescoping, compressing, emphasising and distancing – giving places agency, allowing them to speak and dream, making them seem more rooted, or reinvented, uncanny and hyperreal. Situating art in public places – from graffiti, land art and airport art, to self-conscious landmark placemaking sculpture – creates tableaux to which people can respond. Biennale and triennial exhibitions are popular because they decontextualise places, gifting them with a new sense of origin and relational aesthetic. During exhibitions, places become kinaesthetic, mixing materials and media in a unique interrelationship. Some places retain the permanent signature of the artist, many works are transient, leaving empty spaces behind. Art affects our conceptual authentication of different sites; intangible cultural cues change the resonance of sites and the real and the fake merge into dreamery.

The practice of situating art in specific places could be said to fit with the concept that 'the work must create meaning from or for the place in which it exists' (Smith, 1993, cited in Bishop, 2012, p. 200). Creative places range from destinations with artistic reputations such as St Ives, to national and local galleries; they include sculpture parks, heritage sites, street and public art, national monuments, festivals and exhibitions, popup galleries and events and artists' studios, all of which reflect the increasingly aestheticising gaze (Urry and Larsen, 2011).

Authenticity is involved on multiple levels when encountering art: the physical context, emergent taste and fashion, the objective authenticity of patina, material, era and provenance, the traces of the maker, the value or worthlessness of a

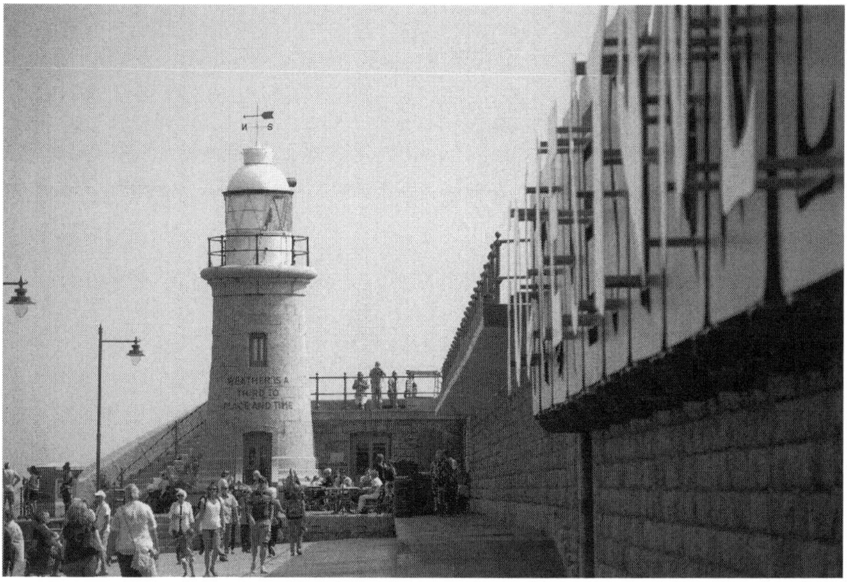

Figure 5.1 Folkestone Triennial 2014, Ian Hamilton Finlay, *Weather is a Third to Space and Time*

Source: Folkestone Creative Quarter

forgery or copy, limited editions, multiple prints or casts. Ai Wei Wei deconstructs these notions in his work, in what he terms anti-nostalgic 'authentic modernism' (Rosbottom, 2015, p. 49). As Locke (2015, p. 155) states,

> Ai is very conscious that markets are full of fakes being sold as originals and that only experts can distinguish between them. The creation of forgeries interests him since the same skills and traditions used to create the originals are used to create modern versions.

Ai Wei Wei deconstructs art and its setting – for example, in *Fragments* he uses ancient temple beams from the Ming and Qing dynasties to make a structure which, when viewed from above, is a map of China. *Fragments* forms a heterotopia, with place available on multiple levels, using original materials which are decontextualised and resituated in a sculpture in a gallery, depicting a country. Deconstructing places through art adds and subtracts the dimensions of context and meaning, which can change our view of the art's value, historicity and authenticity.

The chapter examines the consumption of art and its relationship with authenticity and then explores different spaces of imagination, suggesting that some large-scale, site-specific art can reach beyond real, fake and post-truth to the uncanny, or magically real. Faris (2004, p. 1) defines 'magical realism' as combining 'realism and the fantastic so that the marvelous seems to grow organically within the ordinary, blurring the distinctions between them'. Some types of site-specific art seem to perfectly inhabit the concept of third-order simulacra (Baudrillard, 1981); they are creations for which there is not an existing blueprint. Baudrillard (ibid., p. 166) relates the allegory of the Borges tale, where the cartographers make a map to life-size scale and it covers the Empire. The map wears out and disintegrates, with only a few shreds left decaying in the desert:

> simulation is no longer that of a territory, a referential being or a substance. It is the generation by models of a real without origin or reality: a hyperreal. The territory no longer precedes the map, nor survives it. Henceforth it is the map that precedes the territory – precession of simulacra – it is the map that engenders the territory and if we were to revive the fable today it would be territory whose shreds are slowly rotting across the map.

The chapter continues to explore examples of creative placemaking as a method of storytelling, interweaving history (facts), heritage (story), surreal dreamscapes and simulacra. Placemakers may use culture instrumentally to improve community wellbeing and provide access – for example, by placing public art in a variety of non-designated spaces, ranging from national parks to supermarkets. Finally, the chapter examines how cultural destinations are developed through art and how participative agendas and community ties may lead to the creative authentication of places. Ultimately, the question of who makes the meaning, who pursues it and who discovers it is as full of possibilities as the meaning which is derived.

Places of imagination

Raban (1974) and Selby (2004) identified the 'soft city' of imagination which exists simultaneously with the material world. Cities of imagination are inspired by global, capital, cinematic cities such as London, Paris and New York, as characterised in art, film and literature. Spaces are not simply either soft, or real, but can be *surreal* dreamscapes (Raban, 1974; Donald, 1999; Pile, 2005). The soft city functions at pedestrian level and it is here where creative agency appears to operate and places seem to have a mind of their own. Hyperreal creative places evoke the magical authentication of a topsy-turvy Alice in Wonderland, universe, playing with space. Soft places draw on collective, multi-sensory memories, elements of imagination and the surreal landscape of the subconscious. While art has an intrinsic and commercial value, in the public realm it can also provoke and subvert places, providing a dream spectacle and sense of the fantastical. The artist Cristo is known for wrapping large-scale structures, making them unfamiliar, conjuring the suggestiveness of the hidden interior – for example, Cristo and Jean Claude's *Wrapped Fountain* and *Wrapped*

Medieval Tower, part of the Festival of Two Worlds in Spoleto, a small mountaintop town in Central Italy.

As discussed previously, the search for authenticity encapsulates the desire for ephemeral, magically real, enchanting experiences. Spaces of imagination include the phantasmagoria (Benjamin, 1955), which Jamal and Kim, (2005, p. 66) described as 'a metropolis revolving around artifice, calculated illusion, metronomic spectacle and automated simulation'. The phantasmagoria is a labyrinthine place of myth, dream-symbols and desires, rooted in the nineteenth-century loss of the rural idyll and backdrop of modernity and futurism. Blockbuster exhibitions evoke phantasmagorias, visitors weaving through labyrinths crowded with secret pockets, unexpectedness, interactive effects and lighting. The Alexander McQueen retrospective *Savage Beauty* at the Victoria and Albert Museum in 2015, combined authentic artefacts, recreation and illusion in exhibition rooms resembling the interior of music boxes or cabinets of curiosity. The phantasmagoria can be nightmarish; *Verity* is a stainless steel and bronze statue created by Damien Hirst, a 20.25-metre tall sculpture of a pregnant woman, half her internal organs and foetus displayed, on the pier at the entrance to the harbour in Ilfracombe, Devon, juxtaposing pretty seaside resort with dystopian postmodernity.

As demonstrated in the chapter on gentrification, cities are increasingly commercial and hyperreal, filled with symbols of heritage, shopping, nostalgia and consumption, adding layers of desire to the sense of the phantasmagoria. People are confronted with the material dreamscape (Pile, 2005) of shopping spaces made up of consumer symbols and choice (Benjamin, 1955, cited in Jamal and Kim, 2005, p. 66). When colossal sculptures are added to a commercial landscape, they articulate additional dialogues of playfulness and performance – for example, Claes Oldenburg and Coosje van Brugge stainless steel 12-metre *Dropped Ice Cream Cone* on the top of the Neumarkt Building in Cologne, pointed cone referencing the many steeples in the skyline, an urban magically real fantasy. Or Olderburg's outsized binoculars, which were incorporated into the Frank Gehry design for the Chait Day Mojo office building, providing a spectacle for the myopic gaze of the art-tourist.

Consuming art in places

Experiential authenticity can be induced in the form of the possibilities created by art's consumption. Owning art can be a form of self-expression, a 'cooler authentication' (Selwyn, 1996; Cohen and Cohen, 2012), and appraisal of acquisition, possession and curation, which makes a statement about the constructed self. Bourdieu's (1984) concepts of distinction, taste and status examine the assertion of individuality through cultural capital; the consumption of art is a negotiated process, characteristic of Wang's (1999) theory of constructed authenticity. Grayson Perry elaborates on distinction in *The Vanity of Small Differences* (Perry *et al.*, 2013, p. 12), exploring the visual environment and aesthetics of the home: 'This title comes from the *Narcissism of Small Differences* used by Sigmund

Freud in *Civilisation and its Discontents*, alluding to the fact that we often most passionately defend our uniqueness when differentiating ourselves from those who are nearly the same as us.' This supports the theory of the 'mechanistic authentication' (Lovell, 2013) of the provenance, backstory and setting of artworks as acts of micro-connoisseurship by tourists seeking distinction. Art can give tourists 'private views' and greater insights into places, switching on the gaze. Meaning-making is pluralistic, different perspectives acknowledge something unique about the same piece of work, as Benjamin (1936 [2008], p. 10) illustrates with the example of the statue of Venus.

> A classical statue of Venus, for example, occupied a traditional context for the Greeks who made of it an object of worship… medieval clerics, saw it as a threatening idol. But what both were equally struck by was its singularity or aura.

Wollheim (1970) evaluates the evocative nature of works of art, which he argues possess representational, expressive and physical properties, known as 'significant form'. Wollheim (ibid., p. 39) critiques the hypothesis that an artist has embedded, authentic emotional qualities within works of art, or that objects have the agency to:

> produce a certain state of mind or feeling in the spectator… the work's expressiveness now becomes a purely external feature of it. It is no longer something that we can or might observe, it is something that we infer from what we observe: it has been detached from the object.

The art flâneur is a common figure in art consumption. The gaze of the flâneur is thus not simply mechanistic, but is open and drawn to the possibilities of public art and the hedonic consumption of aesthetics, 'based not on what consumers know to be real but rather on what they desire reality to be' (Hirschman and Holbrook, 1982, p. 94). Tourists read meaning into the attributes and backstories of art, hotly creating intersubjective encounters.

In *The Precession of the Simulacra*, Baudrillard (1981) states that both map and territory are now irrelevant, indicating that the fake, or symbolic, has replaced the real and previous standards of authenticity no longer apply. However, observing reactions to staged authenticity in historic cities, Lovell (2013) suggests that while map and territory are 'shredded', authenticity persists cognitively, emotionally and critically. This is also expressed by Collins-Kreiner (2010, p. 453), who suggests that: 'The point is that although everything may be in a state of flux, we can nonetheless discover structures beneath the surface'. Patterns are conspicuous in the ways in which visitors are predisposed to value art. As a 'sacred journey' (Graburn, 1989) tourism involves travelling to see a place or relic, such as the painting of the *Mona Lisa*; visitors are therefore inclined to reify objects. Sometimes art or installations are in a place that has been invested with meaning. An example of this is the Palaeolithic cave paintings of Lascaux in

France, which are discussed in more detail in Chapter 3, but where we have no way of knowing about the original significance and intention of the work.

It could be said that designated creative places attract flâneurie; galleries and museums are contrived for people-watching and tableaux, the lighting, cafés, glass shop windows and benches directing the gaze. Works of art in designated spaces such as museums and galleries are contextualised and framed by their setting, which acts as a platform or shrine for viewing, enhancing reification in a similar way to sacred objects in religious sites. In its extreme form, the experience of the artwork may become valued and elevated by art tourists to such an extent that the piece is treated as a relic, or object of mass pilgrimage, such as the *Mona Lisa* in The Louvre, or Botticelli's *Venus* in the Uffizi, the gallery becoming a reliquary – auratic, 'named, framed and elevated, enshrined and reproduced' as part of the sacralisation process (MacCannell, 1989). This may hold true even if the painting is found to be smaller than imagined, or photographing crowds seem to devalue the personal experience. Aura is based on the mystery of 'distance' (Benjamin, 1936 [2008], p. 5) and to see an artwork in what Crouch (2002, p. 213) terms the 'close-up' space of practice can seem disappointing, unreal, a fake experience of a real object. But, as in the case of pilgrimage, the desire for an authentic experience may overpower the artificiality of the situation, such as the souvenirs, or the expense, in a 'collective wish fulfilment' (Shackley, 2001, p. 29). However, wish-fulfilment is dependent on aura and aura is dependent on belief. The placebo effect of authenticity may be diminished if we understand that we are viewing a copy of the *Mona Lisa*, not the real thing. To return to the discussion in Chapter 3, Matisse's (1929) painting *The Treachery of Images* is a painting of a pipe, not a pipe. If we are viewing a print of the painting on a wall, a postcard, or a fridge magnet, as discussed in Chapter 9, the representation is even further removed from the original. It is detached from the hand of the maker, no longer a witness to events, or connected to celebrity. Like many artists, Ai Wei Wei subverts the idea of aura; he paints Neolithic ceramics to ensure that they are indistinguishable from recent pottery urns. In the work *Dropping a Han Dynasty Urn* he is photographed and filmed destroying an artefact. He preserves the dust, positing the question of the extent to which Neolithic dust still retains its value or whether a new fusion of aura is vested in the process of the deconstruction, display and reproduction on souvenirs, in photographs and catalogues, as an Ai Wei Wei exhibition piece.

Purposeful or serendipitous encounters with art in public places can encompass existential, intersubjective and emotional elements. Imagination and dreams play a part in making space creative and fluid, as Merrifield (2000, p. 171, emphasis original) describes: 'it flows and collides with other spaces. And these interpretations – many with different temporalities – get superimposed upon one another to create a *present* space.' A key element of travelling is said to involve escaping the everyday, preventing people being reminded of or 'haunted by' their usual roles at home (Crang, 2006, p. 55). According to Smith (2003, p. 28) 'tourism has been viewed traditionally as an escapist phenomenon, one which is based predominantly on elements of fantasy, fictions and myths'. Fabrications and surreality thus draws the tourist gaze. If the map has truly replaced the territory, creative agency makes it seem as if the fantastical dreams of the places themselves are escaping from their everyday existence, expressing themselves through magical realism in a more-than-human world.

Some art creatively authenticates place, adding depth and meaning. The dialogue between art, imagination and place is possibly most directly understood in examples of site-specific commemorative art such as *Blood Swept Lands and Seas of Red* by Paul Cummins and Tom Piper, where 888,246 ceramic poppies filled the Tower of London's moat between 17 July and 11 November 2014, marking 100 years since Britain joined the First World War. The Tower of London setting gave the memorial stronger panoramic impact; it could be seen from many viewpoints, including from boats on the river Thames, the magically real spill of red altering the public perception of the historic monument. The extent of the poppies, each representing a soldier, allowed people to gauge the enormity of the war visually. These elements are also present in Rachel Whiteread's *Judenplatz* Holocaust memorial. Like her other works, it is a cast of a room, in this case a room in a library located in the square. It depicts a seemingly infinite number of library books with the spines facing inward, the inside turned outward, unreadable, symbolising the lost intellects and grief externalised. This is art as simulacrum, designed to provoke thought and pause and stimulate; the art hotly authenticates and memorialises the square and is experienced emotively. The experience can mix sublime place, hyperreal form and exhilaration at seeing a valuable object, which can provoke the state of 'being present' (Eliasson, 2016a), aware of natural details and sensations and a deeply authentic, inchoate, mindful experience, providing a 'punctum' (Barthes, 1980) or pause in the tourist experience.

This section has suggested how art in places can stimulate the cool, cognitive, mechanistic authentication by tourists of provenance and materials, the warmth and exhilaration of consumption, the distinguishing of the authentic self, the hot authentication of the religious encounter with a relic, the imaginative, creative authentication of the traditional, historic site and the magically real discovery of simulacra which defamiliarise and destabilise ordinary places. These affective states may change the atmosphere and reading of places as feelings of magically real authentication evaporate into delight.

Site-specific art

Art and place are often inextricable; site-specific art draws from and informs the character of locations, while remaining both separate and part of them. Discussing his 2016 installation, a suspended waterfall in the gardens of Versailles, Olafur Eliasson (2016a) uses the phrases 'constructed nature' and 'contrived environment' to suggest the way that art, combining with the variables of atmosphere, can manipulate the way we reflect on the meaning of space and place. Describing the growth of sculpture parks in 'spectacular topographies', Langen (2015, p. 9) details the movement from 'white cube' to 'green maze', as large-scale art is positioned in natural environments rather than in traditional exhibition spaces. The resultant green mazes profoundly transform different environments, subverting notions of place. Langen (2015) has extensively researched sculpture parks, including the collections of private patrons such as Patrick McKillen. His park is placed in a vineyard north of Aix-en-Provence, and includes works such as Louise Bourgeois' *Crouching Spider*. Other unusual locations include the Plaine de Marha in the desert in Morocco, containing three works by Hannsjorg Voth made from traditional rammed-earth methods and local materials and relating to the skyscape and transcendence. Further examples of 'contrived nature' include the Museo Subecuatico de arte in Cancun, Mexico, containing sculptures by Jason deCaires Taylor, which are located underwater, allowing entropic processes to reclaim the works as an ecological message. Andrew Rogers references ancient chalk hill carvings and Mayan symbols in his series *Rhythm of Life*, comprising horse and abstract patterns, using stone walls to create the outlines of the shapes (ibid.). Encountering large-scale, monumental simulacra can alter perception, leading to a hyperreal sensation of magnified authenticity, shrinking time, or sense of self transforming from real to magically real.

Hyperreality is intensified in other large-scale individual pieces of site-specific art that involve bringing the unnatural into the remoter natural environment, modifying and extending nature to create a surreal simulacrum. Cristo and Jeanne Claude's magical realist work *The Floating Piers* in Italy's Lake Iseo, staged in June 2016, were made from yellow fabric, held by polyethylene cubes floating on the surface of the water, allowing people to walk on water to the island. In contrast, Richard Long's organic work responds to the environments he has walked in and constructed from found materials, such as slate and wood. The work is then constructed in place, such as the *Connemara Sculpture*, and photographed. The natural world is striking when brought within the urban environment – for example, Olafur Eliasson's *The New York City Waterfalls* cascading from structures in New York Harbor, Brooklyn Bridge. As Eliasson (2016b) states,

> I have tried to work with today's complex notion of public spaces. The Waterfalls appear in the midst of the dense social, environmental, and political tissue that makes up the heart of New York City. They will give people the possibility to reconsider their relationships to these spectacular surroundings, and I hope they will evoke individual experiences and enhance a sense of collectivity.

Place, art and authenticity can grow together. Works can become so situated that the place is synonymous with them, such as the copy of Michelangelo's *David*, placed in the Piazza della Signoria in Florence. Public art projects can use creative authentication to reinvest places with their local cultural capital; Bishop (2012, p. 195) describes heritage sites 'as an evocative formal backdrop for work imbued with historical resonance'. The cities of Hull and Sheffield have lost much of the industrial heritage that gave them their meaning and art is a way of celebrating and expressing the new sense of meaning as they make the transition from de-industrialisation. Creative interpretation of the past gives way to creative authentication, locating the changing essence of places. Anthony Gormley's colossal, steel landmark *The Angel of the North*, located on the site of a coal mine, embodies the leviathan of Northern industrial heritage. The sculpture *Blade* by Nayan Kulkarni, in Hull, depicts an outsized turbine blade that reflects local industry, but is strangely dormant, rather than seemingly propelling the city into a creative future. Some public art reveals the emergent progression of 'real to fake to real' (Lovell, 2013): first, a project such as a piece of public art is developed based on the original city's distinctive heritage; second, it is implemented, scrutinised and possibly resisted by the local community; and, third, it is adopted as part of local cultural capital and 'new heritage'. An example of this process is the Henry Moore sculpture *The Family*, which was intended to give the new town of Harlow, Essex, new meaning, associated with family life and modernity. Gormley's *Another Place* installation is located on the foreshore at Crosby, near Liverpool; tankers pass the 100 sculptures, which are scattered up to 1 kilometre out to sea. It has become part of local mythology and is now known as *The Iron Men of Crosby*; it is an example of an installation resituating the experiential authenticity (Wang, 1999) of place.

City-wide exhibitions remake environments, placing art in unusual places, but also creatively authenticating the 'historical resonance', mentioned earlier by Bishop. *BLOK*, an outdoor sculpture show in Canterbury for 100 days in 2005, was organised by one of the authors, in partnership with English Heritage and the University of the Creative Arts. The show was intended to create a historic-creative space, relating large-scale sculptures to the heritage environment, placing the uncanny in the everyday, using World Heritage sites including Canterbury Cathedral and St Augustine's Abbey. The juxtaposition of monumental city walls and modernist sculptures in steel or, for example, the water sculptures of Bill Pye, created a magical realist prompt. The sculptures were used by the public, who accepted them as new permutations of place, climbing on them and photographing them; the works were rapidly absorbed into the *mis-en-scène* as part of the heritage experience (Canterbury City Council, 2005). Pablo Bronstein's *Beach Hut in the Style of Nicolas Hawksmoor* was placed among a row of ice-cream-coloured huts at Folkestone Triennial, enabling people to gaze again at a common seaside sight.

These exhibitions are transient, the only opportunity to see the art in specific locations, although some commissions are retained in place, becoming integral. An example of retained art is *Baby Things* by Tracey Emin, a Folkestone Triennial work which depicted unsettlingly bronze models of delicate baby clothes and booties, under a bench and draped on railings, embodying the high number of teenaged pregnancies in the town.

Creative placemaking

To an extent, creative places can be made, designed in a utopic, constructed, conceptual manner (Lefebvre, 1991; De Certeau, 1984). Creative placemakers

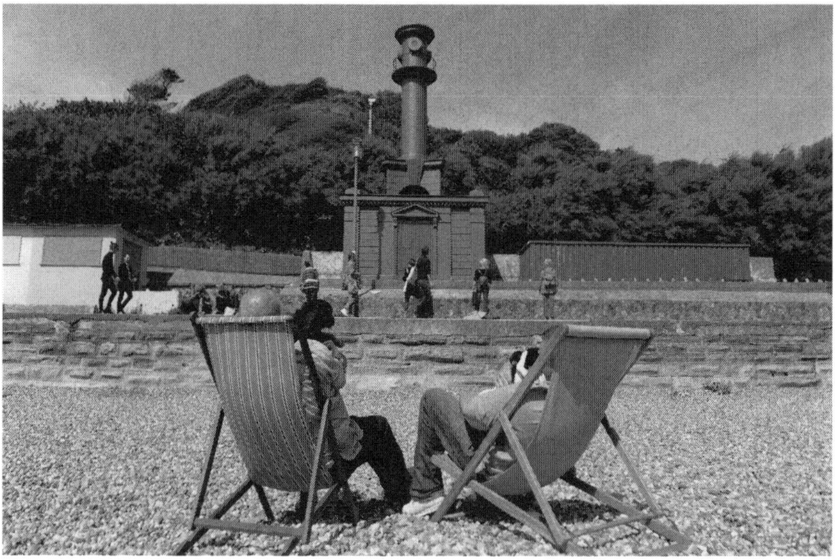

Figure 5.2 Folkestone Triennial, Pablo Bronstein, *Beach Hut in the Style of Nicolas Hawksmoor*

Source: Folkestone Creative Foundation

are described by Richards and Wilson, (2007, p. 15) as 'imagineers', the term Fjellman (1992) used to describe the planners of Walt Disney World in California. In this section, it is argued that, to be authentic, creative placemaking should be participative and truly located in the historical and current community. One of the most striking early examples of cultural engineering can be seen in the Federal Art Project (FAP) (1935–43) launched by President Roosevelt during the Depression of the 1930s in America as part of the Works Projects Administration (Willett, in Baskerville and Willett, 1985). It was intended to employ people not simply in a basic capacity, or by awarding the dole of the time, but by enabling them to do the job they had trained for. It had an impact on the position of New York in the art world and abstract expressionists such as Pollock, Rothko and Kline were employed in the projects. Willett argues that the FAP tended to be selective and conservative in subject matter – for example, social realism was labelled un-American and communistic – and 6,000 artists were funded in its first year, with projects, materials and equipment. One of the most well-known outcomes was the use of murals on government buildings, emulating a similar project in Mexico in the 1920s, produced by artists such as Orozco, Siqueiros and Rivera. The use of large-scale murals by artists such as Thomas Hart Benton, underpinned the concept that the FAP was 'predicated on the belief that art should not remain the preserve of the privileged class' and the choice of buildings in the public realm meant the work was labelled 'post office art'. Public Art Fund and ArtPlace America invest across contemporary America in artist-led placemaking projects, siting art in non-designated spaces.

Like aspects of gentrification, creative placemaking can involve a top-down method of planning, constructing new authenticities, meaning and identity. 'With the disappearance of local manufacturing industries and periodic crises in government and finance, culture is more and more the business of cities – the basis of their tourist attractions and their unique, competitive edge' (Zukin, 1996, p. 2). Richards and Wilson (2007, p. 15) refer to the 'creative turn' or enculturation of the economy, where the creative industries are fostered in a way designed to attract the 'creative class' (Florida, 2002; Landry, 2000; Maitland, 2007; Zukin, 1995) and creative tourism. As described in Chapter 4, the publication of *The Creative Class* (Florida, 2002) marked a recognition and incorporation of strategic creative industry development (Bishop, 2012; Flew, 2012; Florida, 2002; Landry, 200); nurturing a creative workforce was adopted as an economic strategy by the UK's New Labour government (Bishop, 2012; Chakrabortty, 2011). The flourishing experience economy (Pine and Gilmore, 1999) also became central to the resurgence of many places. This led to the hyperproduction of cultural symbols, such as museums, galleries, theatres, public art and architecture (Richards, and Wilson, 2007; Hughes, 1998), which were designed to decrease the forbidding nature of many traditional art places and increase accessibility. The UK Arts Council's Creative Places and the Department for Culture, Media and Sport (DCMS) Great Places Scheme both focus on community-led art development in place and it could be said that the search for authenticity by tourists is fed by constructed narratives which amplify the cultural pedigree of places through art. One of the authors was involved in East Kent's bid to be UK City of Culture in 2017, bringing heritage, creative and tourism industries together 'in the same room', resulting in several interdisciplinary projects, including Maze, a collaboration between Turner Contemporary and Jasmin Vardimon Dance Company. The bid provided an awareness of a broader spectrum of arts venues outside the gallery, from castles to museums, creating opportunities to creatively authenticate places in more imaginative ways.

Cultural quarters usually include a creative cluster of businesses – for example, Vienna's Museum Quartier, containing the Mumok (Museum of Modern Art), the Leopold Museum, creative spaces and performance/exhibition venues. Roodhouse (2006, p. 54) suggests that they are designed to reinforce the artistic identity of the city and values of the region. Creative clusters based on the geographical grouping of assets have also been adopted as a strategy for consolidating cultural development, to promote brand identity, reuse old buildings, reinforce value chains, encourage entrepreneurialism and stimulate microbusinesses and cultural diversity for those marginalised by the art world (Mommaas, 2004, in Flew, 2012, p. 149). As suggested earlier by Richards, some cities invest in cultural attractions, but smaller destinations use their urban space and street scene to develop cultural character. Creative clusters can also occur in less-developed areas of cities or seaside resorts, due to the availability of affordable rent and studio space – for example, Kent's Creative Coast, which is grouped around the anchor towns of Whitstable, Folkestone and Margate and their galleries and events; Whitstable Biennale and Folkestone Triennial are linked through the public-sector initiative Culture Kent. Cultural engineering by patrons, including Roger de Haan's Creative Foundation, provides infrastructure for the artistic community, involving the purchase of shops in the winding, narrow Old High Street in Folkestone, which are used for studio and gallery spaces. This, in turn, has attracted new businesses such as RockSalt, a Michelin-starred restaurant in the harbour. The modern design of Turner Contemporary contrasts with the seaside vernacular in the context of the seaside town of Margate, yet the Jerwood Gallery in Hastings, located by a working fishing beach, is designed to be in place, mirroring the tall, dark fishermen's huts. The Turner Contemporary gallery is a National Portfolio Organisation, using art instrumentally to reinforce social capital; art tourists and artists gravitate to these attractions, which makes a statement about the cultural social values of the destination, offering potential hope, opportunity and inward investment to the community.

Roodhouse (2006, p. 54) argues that some cultural quarters could be potentially over-dependent on practitioners and networks, which could mean the loss of the creative industry after 'a generation' of people have moved on. Ward (2016, p. 2) describes 'artful brands' which can be utilised for placemaking, without truly supporting the artists, who live a parlous existence unless meaningful infrastructure is in place; for example, recent changes at the Chocolate Factory, which houses artists' studios in Wood Green Cultural Quarter in London, have meant that many artists have relocated. While some places like Margate and Folkestone have the resources to ensure that art and community are inter-reflective and linked, in other incidences cultural placemaking does not always successfully mitigate social issues; the artistic and cultural tourist communities can sit side by side unintegrated, with the sensation at the destination of turning right for cupcakes and left for despair. This makes the experience unsettling, the authenticity of place is evidently not located in the art, but elsewhere in the gritty town and the consumption of art is uneasily refracted through a prism of deprivation.

Achieving an authentic sense of creative place involves a holistic, instrumental approach, drawing on Public Health funding for lifelong learning and wellbeing projects which were aspects of the Transformation Agenda designed to cut public sector costs in the UK. Truly authentic projects could be argued to be thoroughly participatory, enacting the principles of inclusion and access. However, discussing 'outsourcing authenticity', Bishop (2012, p. 219) compares the use of local people to film directors choosing actors to play roles, and suggests that the artistic and local community do not always truly mix during the project (ibid., p. 199). Participative art could be argued to include the work of Spencer Tunick, whose installations involve large-scale nude photographs of naked people – for example, 3,200 participants painted blue in Kingston Upon Hull for the *Sea of Hull* work, part of the 2017 City of Culture celebrations, a modern reworking of the community's maritime relationship.

Remaking creative places may invert gentrification, reinjecting the art (albeit with an appraised value-system) back into abandoned, 'gritty' (Zukin, 2010a) communities, drawing on the cycle of real to fake to real. The ceramicist and social activist Theaster Gates turned an abandoned house in Chicago into an artwork, *The Archive*, containing repurposed books from a shut-down bookstore, added a soul food kitchen and opened it to the public. The Rebuild Foundation, 'a slum take back project' (Jones, 2016), has continued to transform a series of South Side Chicago buildings, the Dorchester Projects, into cultural centres and archives for music. Gates also bought a flooded bank for a dollar and sold bonds using its urinals (which are inscribed 'In Art We Trust') to begin renovations. Architecture and Design Collective Assemble won the Turner Prize in 2015 for their work in Toxteth in Liverpool. These places then gained an extra dimension, becoming objectified for the visitor gaze as artworks with implicit grit; their surroundings suggesting a greater depth of creative authentication because they are reworked from the inside-out – staged backstages, but reflecting more accurately what is there.

Conclusion

Examples provided within this chapter suggest that places are co-created by artists, placemakers and tourists, all of whom have different approaches and intentions but all will have the potential to creatively authenticate places. Creative authentication reanimates places, enhancing their aura and meaning, from post-industrial sites, or seaside resorts to seemingly placeless locations such as airports, or new garden villages.

The chapter has contemplated the potential for the spectacle of art simulacra in outdoor and unexpected contexts to conjure up magically real dialogues between tourists and places. Academic debates about authenticity, art and creative place-making have often been focused on social value, wellbeing and interdisciplinary work, but the works have their own existence and life-stories and the conversation and memories which they generate are where the authentic community involvement truly resides. The creative interpretation and mechanistic authentication of places is apparent in vernacular materials and references, a shift of what is there, or perhaps, using the example of post-industrial heritage, or commemoration, a rediscovery of what has been lost. Places are defamiliarised and softened by art, becoming distant, more auratic, creating a link between the observances of flâneurie, and the longer pause to note the poetic everyday. Site-specific installations offer a situationist, non-representational, embodied encounter, stimulating the imagination, offering the deeply evocative, poignant, uncanny, superheated evaporatingly magical authentication of places.

The chapter began by discussing the words of Georgia O'Keeffe, an artist who drew heavily on images of place and whose work became inseparable from its inspiration, the Sonoma Desert in New Mexico. She became increasingly introverted, living quietly at Ghost Ranch, New Mexico, the place, art and person becoming indistinguishable. One of the authors tried to meet her at an exhibition in Santa Fe, but the car broke down in the desert and the exhibition was missed. Missing Georgia was appropriate. In the BBC documentary *Imagine Georgia O'Keeffe: By Myself* (2016), she expressed a desire not just for personal privacy, but also to keep New Mexico secret and hidden, a backstage zone only she could access. In the paradox of art and place, if places are discovered to be truly creative, commercial elements drive the prices up and the artists out; the very act of consumption could, in this light, be interpreted as an act of inauthentication. As the wheel of authenticity turns again, a downturn in the economy may force a 'new originality' as artists return as a more organic part of creative places.

6 The seaside resort

From piers and promenades to nostalgic authenticity

According to Brodie and Winter (2007, p. 1), 'the seaside resort is a special place' and this distinctiveness is the essence of its singularity, incorporating the smells, taste, sounds and sights of the coast. Created ostensibly for leisure, resorts exhibit many key features arising from a unique set of circumstances. In part, this reflects their physical geography and marginal location (Ryan, 1997, 2002; Shields, 1990; Walton, 2000), but they also have a distinct built environment and urban morphology resulting from economic, political and technological forces (Jeans, 1990; Pearce, 1995; Robinson, 1976; Shaw and Williams, 1997); and they are socially and culturally constructed spaces (Urry, 2002, Shaw and Williams, 2004; Walton, 2000). Although resorts are thus human constructions, artificial environments resulting from the transformation of undeveloped coastline or pre-existing ports or fishing villages, their uniqueness affords a vivid contrast to the humdrum conditions of people's everyday lives and an essential escapist destination for those in search of the 'authentic Other'.

Resorts began to develop in their present form from the end of the eighteenth century and had their origins in the search for improved health, fashion and, as Urry (2002) would argue, new places on which to gaze. Their growth was enabled by many of the developments of industrialisation and stimulated by the need to escape the increasingly polluted urban centres. They provided a variety of leisure opportunities for all social classes, through day trips as well as longer holidays, and provided substantial opportunities for commercialism and the involvement of municipal government. For almost a century, from the 1870s onwards, resorts in Britain and other parts of northern Europe enjoyed substantial growth and popularity and it is during this period that this most distinctive and instantly recognisable place, what might be called the authentic resort, was fashioned. Resorts, however, are dynamic places and their environments have been affected by economic and cultural change, especially in recent decades which have seen various economic restructuring and regeneration initiatives in response to economic decline (Agarwal, 1999, 2002; Cooper, 1990; Gale, 2005; Shaw and Williams, 1997, 2004; Walton and Browne, 2010). Nevertheless, despite attempts to change their character and focus, there are sufficient elements that enable an air of authenticity to endure, such that many resorts display a range of both modern and relic features.

In a similar way to historic cities, resorts capitalise on a sense of their originality, which, as the chapter will explore, is based on the concept of a heyday forged in more recent history. Ferry (2009, p. 124) described how, at the beginning of the twenty-first century 'there are real signs of a renewed interest in the seaside as a holiday destination' and for many people its continuing attraction and authentic appeal is linked to nostalgia, especially associated with memories of childhood seaside holidays (Jarratt and Gammon, 2016). Jarratt and Gammon have described how resorts such as Morecambe draw on nostalgia, often inspired by childhood visits. Nostalgia is an essential part of seaside authenticity and 'such affection for the past' is fuelled by an almost continual outpouring of popular books describing and illustrating past images and memories. Some simply feature individual resorts, while others, such as Lycett Green's (2011) *Seaside Resorts* and Freeman's (2015) *Beside the Sea: Britain's Lost Seaside Heritage*, describe a variety of different ones. But all are richly illustrated with photographs of a bygone age. Authenticity has been argued to emerge, but it can also retreat, and resorts have experienced a decline in more recent years – although some have arguably had a renaissance as cultural destinations, as discussed further in Chapter 5. This chapter explores the combination of a diffuse sense of loss of the golden era of resorts, the childhood visits and their manifestation in a sense of 'warm authentication' (adapted from Selwyn, 1996, and Cohen and Cohen, 2012) which is also central to Chapter 7. Their aura, the 'seaside idyll', continues to affect the way in which we see seaside resorts and the way in which we behave. In a similar way to theme parks, the warm authentication of resorts involves searching for signs of representation and commodification, such as saucy postcards and tropes of childhood, such as sticks of rock, buckets and spades. These may seem artificial; however, they are central to the notion of the authentic resort. The seaside also offers a range of experiential authenticity (Wang, 1999), from the existentiality of experiencing a liminal place through to the spirituality of light and openness.

This chapter assesses the extent to which such places might be regarded as authentic, considering the characteristics of the modern seaside landscape, and examines the various socio-cultural, economic and political processes that have led to its creation and continued transformation. It focuses on the British seaside resort, partly because to include broader coverage would result in a more superficial account and also because several authorities have argued that seaside holidays were a British invention (Ferry, 2009; Walton, 2013). However, many cultures have their own versions of seaside resorts, from Coney Island to San Remo, and the chapter therefore has international resonance.

The historical development of seaside resorts

As outlined above, the authenticity of seaside resorts is linked to their historical development and associated nostalgia and thus it would seem appropriate to begin with a brief coverage of this history. The historical development of resorts is well documented (Hern, 1967; Shaw and Williams, 1997; Urry, 2002; Walvin,

1978; Walton, 1983, 2000) and involves a social, cultural and economic history linked to fashion, social class values, lifestyle choices, health and various economic consequences resulting from the industrial revolution, such as increased wealth, improved transport facilities and the commercialisation of leisure. As mentioned above, resorts developed as distinct tourist destinations from the end of the eighteenth century, but they may have their antecedents in the earlier development of the spas, which became increasingly fashionable among the upper classes and landed gentry after about 1660 (Cosgrave and Jackson, 1972; Patmore, 1968; Towner, 1996). The most fashionable spas were located at relatively inaccessible places to make it costly, and therefore difficult, for anyone other than the very rich to visit. It was this social exclusivity that was also one of the key factors in the early development of the seaside resort because, as increasing numbers of the new middle class also began to patronise the spas towards the end of the eighteenth century, the upper classes began to visit the seaside to avoid these 'persons of inferior quality'. The fashionability of the seaside, as with the spas, was aided by the proclaimed health properties of seawater (bathing as well as drinking, although the fashion for the latter was relatively short lived!) and by royal patronage, with regular visits by the Prince of Wales (later to become George III) to Brighton after 1783. After 1800, it became more fashionable to be seen at the seaside than at spas, but before too long the middle classes were also visiting seaside resorts, with the result that the upper classes retreated to the more exclusive and less geographically accessible places such as Bognor, Burton-St-Leonards and Worthing.

According to Lowerson and Myerscough (1977), the English seaside resort in its modern form first became established in the 1840s and many of the classic resorts such as Brighton and Blackpool were originally middle-class developments. Although people could travel to some resorts by boat, it was not until the development of the railway network from the 1840s onwards that a convenient form of transport was available to transport substantial numbers of visitors from towns and cities to the coast; this facility was further enhanced from the 1870s onwards with the introduction of cheap excursions. In addition to being a key factor in the growth of resorts, the railway also had an important effect on their spatial structure, an aspect that will be discussed in the next section.

The period 1870–1930 is sometimes described as the 'heyday of the resorts', the phrase, listed in various publications (e.g. Cosgrove and Jackson, 1972; Ferry, 2009), evoking the very essence of nostalgia, and it was during this time that what might be regarded as the authentic seaside resort was established. The age became engrained as the essence of the place and resorts continue to draw on this golden era to underpin its character; and it is this search for the authentic resort which underpins many seaside visits. Increased leisure time, better wages and cheap rail fares enabled many millions of people to flock to their nearest seaside and they became more and more the playground of the working classes. In addition to day trips, increasing numbers of people were also taking annual holidays at the seaside. At the end of the nineteenth century, Great Yarmouth was receiving up to 800,000 visitors a year, Margate 700,000, Scarborough 450,000 and

Blackpool possibly about 2 million (Walton, 1983). According to Myerscough (1974), by 1911, 55 per cent of people in England and Wales took at least one trip to the seaside; in order to cater for such large numbers of visitors, the resorts grew from relatively small settlements to quite large towns. For example, whereas in 1871 the population of Blackpool was approximately 6,000, by 1930 it had grown to over 100,000. Resorts that attracted substantial numbers of working-class visitors were those that could provide cheap accommodation. This was certainly true of Blackpool, where land ownership was very fragmented and thus – unlike some other resorts, such as Southport – there were no large landowners who could prevent the development of low-quality boarding houses (Urry, 2002). Brighton was another such resort where, even as early as 1870, there were as many as 800 boarding houses (Lowerson and Myerscough, 1977).

The seaside provided a variety of leisure for all social groups, who could enjoy the fresh air, sand and sea bathing and the natural, authentic elements of the coast; but, as the resorts developed, municipal investment provided a variety of built facilities such as promenades, piers, parks and swimming pools and increasing commercialisation brought pubs, theatres, dance halls and funfairs. However, the cheap rail links which brought the working classes to the seaside did not produce much social mixing. Just as the upper classes had gone to great lengths to avoid the 'despised middle classes', so too various attempts were made by middle-class groups to distance themselves from their perceived inferiors. They too travelled in increasing numbers abroad and they also patronised more remote and more exclusive resorts. In addition to poor accessibility, often through poor rail links, some resorts, as Cosgrove and Jackson (1972) state, also retained an upper- and middle-class clientele by refusing to provide the same amenities that were present in the popular resorts (e.g. St Annes, near Blackpool; Cliftonville, near Margate; Kemp Town, near Brighton; and Frinton, near Clacton). Finally, the different social classes also avoided each other by visiting the resorts at different times of the year. Lowerson and Myerscough (1977) have shown how, in Brighton, distinct seasons operated. Excursionists visited in July and August, family holidaymakers in spring and late summer and the wealthier classes in the autumn, which was regarded as the fashionable season.

The growth of British seaside resorts was temporarily halted by the Second World War, but, thereafter, the growth swelled again, 'being driven on by the deferred demand of the war years as Britain returned to peacetime consumption patterns' (Williams and Shaw, 1997). However, this renewed growth was short lived, for the 1960s witnessed the start of a major decline in the fortunes of British resorts as people began to desert them for newer resorts in the Mediterranean and elsewhere, where good weather could be guaranteed and where modern hotels could provide far better facilities. This decline has produced a variety of responses which have had various consequences for the physical environment of resorts, one being how to cope with a historical legacy of old hotels, theatres, piers, funfairs, arcades, parks and gardens, designed originally for Victorian and Edwardian visitors, that have clearly seen better days and which appear outmoded and no longer relevant to the lifestyles and cultural aspirations of the potential visitor of the early twenty-first century (Bull and Hayler, 2011). Nevertheless, despite their decline, they still attract substantial numbers of visitors. For example, in 2015 there were almost 25 million seaside holidays (involving at least one overnight stay) in Great Britain (Great Britain Tourism Survey, 2015), in addition to 133 million day visits to the seaside (Great Britain Day Visits Survey, 2015).

The character of resorts

As mentioned at the outset of this chapter, the seaside resort represents a unique environment and 'novel urban form' (Gray, 2006, p. 7) with a specific spatial structure and distinctive architecture which makes them original. Although the precise nature of different resorts will vary due to their individual physical geography, such as the presence or absence of beaches, headlands, cliffs, lagoons and dunes, the basic structural features are frequently repeated (Patmore, 1983; Pearce, 1995). The spatial form of resorts is largely determined by the marginal location of being beside the sea (Ryan, 1997), with many of the most prominent and characteristic features running parallel to the waterfront. This clear linear pattern typically consists of the beach, a promenade, road or railway and a first line of buildings comprising the densest and most expensive forms of accommodation and a core of tourist-related shops, bars and restaurants (Pearce, 1995; Brodie and Winter, 2007). In addition, there is also a gradation of accommodation in terms of height, density and price behind the seafront, with high-priced hotels and high-rise apartments giving way to smaller and less expensive hotels, villas and guest houses, which eventually merge with residential accommodation and other urban functions (Pearce, 1995). If visitors continued to travel to the resort by rail, they tended to stay relatively localised, with all their activities and amusements being restricted to the beach, pier and promenade. This resulted in resorts continuing to remain relatively compact. In addition, the railway also influenced the location of accommodation and other facilities, with many resorts acquiring a characteristic 'T-shape', 'the stem of the T being the main street leading from the railway station to the promenade' (Robinson, 1976, p. 162).

The beach

Perhaps the single most important feature of the physical structure of seaside resorts, one that is central to any notion of authenticity, is the beach. While the seaside resort is a novel environment, it is the beach that is the unique element 'at the core of seaside resort identity' (Walton, 2000, p. 19). As Ferry (2009, p. 77) points out:

> To sit or lie in the salty breeze was as wonderfully unfamiliar as the feeling of sand underfoot, turning simple pleasures into something out of the ordinary. So many things that we associate with the British seaside belong to the beach: donkeys and deckchairs, rock pools and sandcastles, Punch and Judy, and buckets and spades.

The beach constitutes the edge, the boundary with the sea (and other countries), and the free and open leisure space that can provide enjoyment for all ages; and its importance has attracted a range of academic interest (Shields, 1990; Lencek and Bosker, 1999; Walton, 2000; Ryan, 2002; Franklin, 2003; Ferry, 2009).

The combination of light, smell, sea and shore can combine to create an effect of experiential authenticity which attracted painters such as J.M.W. Turner to Margate. The beach is a place of natural attitudes and, in contrast, performance. Beaches vary considerably in terms of their length, width, slope characteristics, physical composition (e.g. type of sand and shingle) and their accessibility and such variation will influence the nature of activities pursued. In addition to sea bathing, swimming and paddling in the adjacent sea, which were the initial attractions, beaches are used for passive activities such as relaxing, reading, watching, sunbathing and picnicking, as well as a host of more active pursuits such as walking and walking the dog, jogging, informal games and sports, and the more recent active pursuits of dune buggy riding, wind surfing and kite surfing. Different ages pursue different activities; for some, the beach is also a place for wildlife study and for others a place to sit in a deck chair and watch the world go by as they breathe the sea air and possibly enjoy an ice-cream, while children may be content to paddle in the sea, make sand castles or search for unfamiliar wildlife in rock pools. Despite the growth in recent times of more active pursuits, much of what takes place on the beach remains passive and informal. As Page and Connell (2010, p. 406) point out 'the coast [beach] is… unlike other environments, as people often go there simply to do nothing other than be reflective, physically inactive or simply to "chill out"'. Ryan (2002, p. 166) considers the 'beach experience' and suggests one way of analysing this is by considering the motivations of the beach user. He notes the work of Blackman *et al.* (cited in ibid.), who noted ten basic motives and benefits sought: 'diversion', 'family togetherness', 'experiencing nature', 'outdoor activities', 'solitude', 'physical fitness', affiliation', 'beach activities', 'self-awareness' and 'status'. These various motives were subsequently combined to form three factors: the first being linked to sociability and being with other people, the second essentially being about a safe environment and a place for children and the third composed of being close to the sea, being able to relax and to enjoy the natural scenery. The beach is thus a varied and complex environment that continues to be the key attraction of the seaside resort as a place for tourism.

In addition to being a physical phenomenon, the beach has also been culturally constructed as a specific environment for leisure and pleasure and thus exhibits elements of Wang's (1999) constructed authenticity. Shields (1990) views the beach as a 'liminal' zone where visitors can relax and behave in ways that are different from the patterns and rhythms of everyday life that are governed by accepted mores and codes of behaviour. Resorts began their tourism life as medicinal places; in the mid-nineteenth century they became places of pleasure and fun, which is a reputation they still retain, as places of wellbeing. This sense of wellbeing infuses behaviour and performances of being by the sea. As a liminal zone and environment for playfulness the beach affords opportunities for experiences that are clearly consistent with Wang's existential authenticity (ibid.). In fact, as Ryan (2002, pp. 156–7) points out, 'beaches are one of the few areas where it is socially tolerated for adults to have fun as distinct from leisure'. Urry (2002, p. 29) comments that this liminal zone was also characterised by carnival 'as the beach became noisy and crowded, full of unpredictable social mixing, and involving the inversion of social hierarchies and moral codes'. Carnival not only implies fun but in its lewdness and chaotic, unruly character it also represents a challenge to the ordered world of norms and ethics (Ryan, 2002). While the contemporary beach has lost much of its rowdy, carnivalesque character, the reversal of norms is still apparent. The body is on display (Obrador Pons, 2003) and the beach is one of the few places where people can appear scantily dressed (in what is euphemistically described as 'beachwear') or even with no clothing at all (e.g. the nudist beach). Furthermore, the beach is also a liminal space for children, where they can investigate rock pools and collect seaweed, 'getting wet and muddy in the process in ways which would not have been sanctioned at home' (Walton, 2000, p. 95).

The salt in the air and sight of the expanse of beach also evokes strong feelings of nostalgia, and people's recollections of the seaside are often remembered in terms of childhood visits where the beach was the ultimate attraction, bucket and spade holidays being fondly remembered. Jarratt and Gammon (2016) contend that nostalgia can be considered as an idealised and selective representation of the past and something that reflects dissatisfaction with the present. It offers a refuge that is 'in stark contrast to contemporary experiences of globalization, fast-paced technological innovation and a world which appears to have sped up' (ibid., p. 126). However, Jarratt and Gammon also point out that 'nostalgia does not just stem from a sense of discontent with the present, for it can also be a reaction to an uncertain or perhaps limited future' (especially for older people) and for many it can draw them back to 'an unchanging past that offers both solace and safety'. 'Whilst nostalgia is not restricted to any life stage or part of the past, childhood is a common theme within nostalgic recollections' (ibid.). Such experiences are like those of Jonathon Dimbleby (2010, p. 140) who, for example, tells of his childhood visits to West Wittering and recalls that 'when the tide was in, we braved the breakers and when it was out, we built sandcastles and paddled in the lagoons looking for crabs and sea urchins'. Harrold 'Dickie' Bird (2010, p. 76) talks of his 'love affair' with Scarborough which started in his childhood and recalls, among other things, games of beach cricket, together with other traditional sports. Walton (2000, p. 3) cites the example of the many children's books that 'celebrate sea and sand, bucket and spade, crabs and starfish, coves,… ice cream, Punch and Judy and other ostensibly innocent forms of beach entertainment', which thus reinforce the attraction of the beach. He believes this to be 'an innocent and timeless vision of the seaside which the post-war generation embroidered from its own memories of childhood holidays' and this vision is also repeated by Lycett Green (2011, p. 12), who recalls 'the memory of childhood holidays; memories which we in turn pass to our children and grandchildren'. These no doubt help to reinforce the perceived attractions of the seaside which continue to romanticise the memory of family visits.

The built environment and cultural constructions

While the seaside is shaped by the natural phenomena of beach, sea and associated wildlife, resorts are also characterised by a vernacular seaside architecture which affects their essential, authentic seaside sense of place or 'seasideness' (Jarratt, 2015). In addition to distinctive architecture, there are other cultural artefacts and practices associated with entertaining visitors that help to locate resorts. The distinctive architecture is one of the most obvious features of the seaside resort and is examined in considerable detail in Gray's (2006) text *Designing the Seaside: Architecture, Society and Nature*. The seaside possesses some unique architectural forms such as piers, promenades, beach huts (Figure 6.1) and open-air swimming pools and lidos, as well as those which, although not unique, are still a key part of the instantly recognisable fabric such as funfairs, amusement arcades, theatres and various types of hotels and other forms of accommodation, designed 'in response to shifting patterns of holiday making, social relations and attitudes to nature' (ibid., p. 280).

These elements of the built environment are designed for staging, for seeing and being seen; however, they are perceived as the authentic seaside architectural rhetoric. Gray describes how 'individual resorts often have a particular architectural feel and design character because of the place-specific dimensions of buildings for sleeping by the sea' such as the Regency terraces of Brighton and Hove. Another development that created a further unique form was the

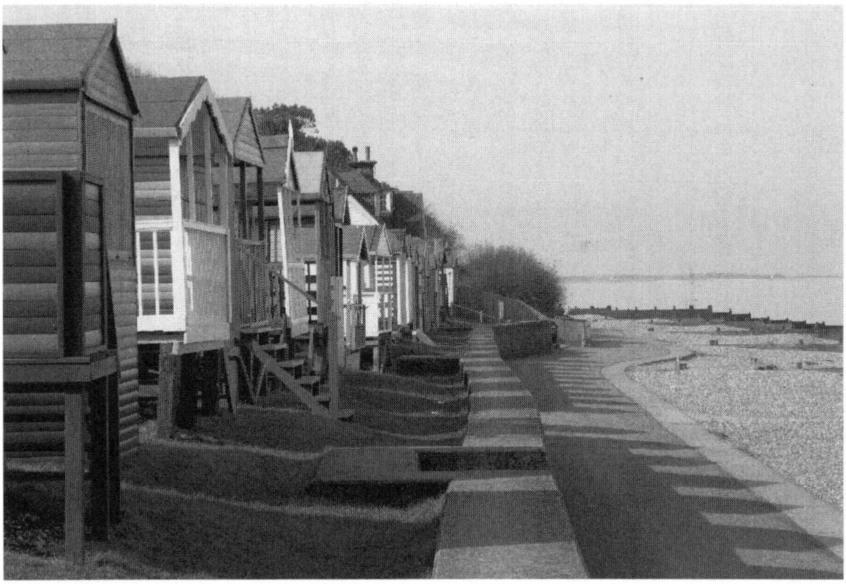

Figure 6.1 Tankerton Slopes beach huts
Source: Canterbury City Council

holiday camp, popularised by Butlin who opened his first camp in Skegness in 1936, to be followed by various other camps throughout Britain (see Ward and Hardy, 1986).

While the beach is natural (although shaped and maintained), the other structures surrounding it are designed for the performance of authentic seaside behaviour. Piers and promenades are uniquely associated with seaside resorts. Ferry (2009, p. 101) states that 'by the end of the 1870s every self-respecting seaside resort had a pier', with 85 having being built around the coasts of England and Wales (Walton, 2000, p. 113). Wills and Phillips (2014, p. 1) believe the seaside pier to be 'the most iconic symbol of the British holiday resort' and that for many 'it is the epitome of excursions to the seaside'. Piers were originally built as mooring facilities to allow the early paddle steamers to bring tourists to certain resorts, but with the advent of the railways piers gradually changed from being landing stages to promenades where people could 'enjoy the pure sea breezes without the danger or difficulty of going out in an open boat' (personal quotation, cited in Brodie and Winter, 2007, p. 136). Later in the nineteenth century 'developers realised that they could house a wide range of visitor attractions' which included kiosks selling sweets, ice-creams, chips and various ephemera; various amusements ranging from simple Mutoscopes (or 'What the Butler Saw' machines) to whole, albeit small, funfairs (ibid., p. 137). Some piers included theatres 'where some of the country's best-loved entertainers performed "at the end of the pier"' (ibid.) and, at some of the larger resorts, theatres were incorporated within much larger buildings or pavilions. Weston-Super-Mare's Grand Pier Pavilion, for example, could seat 2,000 people (Ferry, 2009, p. 103). As well as theatre and music, some piers also introduced dance-floors; Morecambe's Central Pier, in 1906, offered dancing all day (Walton, 2000, p. 105). The piers are highly vulnerable and images of their destruction through fire seems like a metaphor for the decline of resorts in the gaze. Ruined piers in Brighton, Eastbourne or Southend, as discussed by Lovell (2013), are often left in place for many years until the money to rebuild them is in place. Some resorts have now implemented vertical piers, such as the Spinnaker Tower in Portsmouth and the Brighton's British Airways i360.

In addition to piers and their opportunities for promenading, the promenade itself running parallel to the beach was also a key feature of resorts. From the later Victorian period 'attention focused upon the promenade and its transformation into a thoroughfare that was different from any to be found inland' (Ferry, 2009, p. 89). During this time, new walkways were created, gardens were planted, ornamental railings erected, public seats provided and bandstands established. In addition to the opportunities for passive leisure, promenades also provided facilities for sports and games (e.g. tennis courts, pitch and put, crazy golf) and 'when night fell, the power of electricity was harnessed to turn the seafront into a twinkling version of its daytime self' (ibid.). Promenades, esplanades and marine parades constitute further instantly recognisable features of seaside resorts, what Ferry (2009) refers to as that special place, the 'seafront', a place to take the air, see and be seen, saunter with family and friends and to join in the jostle and bustle of the seaside crowds. In addition, funfairs and amusement arcades provided further attractions and novelty to the seafront.

The seaside has also been associated with the extraordinary and exotic such as Blackpool Tower and Brighton's Royal Pavilion, designed for the Prince Regent (later George IV), and West Pier. Both Brighton's structures were oriental in style with Venetian forms being evident at British seaside resorts such as Great Yarmouth, Ramsgate and Southport (Gray, 2006). A further example of the exotic involved the use of tropical vegetation, especially the palm. Artificial palms were used in many resorts, especially in the north of England, although real palms were often introduced in the warmer climes of the south west resorts in what came to be known as the English Riviera.

One feature of the built environment, the seaside theatre, is also linked to the broader cultural form of live entertainment. Live entertainment has been an established feature of the English seaside resort since the nineteenth century, with early examples including itinerant minstrels and the more organised and sophisticated Pierrots who, by the mid-1890s, 'had come to characterise beach music and entertainment' (Walvin, 1978, p. 78). They constituted the first real commercialised live entertainment and played a prominent role in establishing 'the seaside resort as a space for early "mass" entertainment in the evolution of the tourism phenomenon' (Morgan and Pritchard, 1999, p. 2). However, despite their popularity, out of doors entertainment soon came to be challenged by a need for indoor live entertainment that warranted major investment in bandstands, piers, theatres, music-halls and winter gardens. These pleasure palaces and pavilions, together with the 'big productions' within them, became symbolic of the late Victorian and Edwardian seaside and created not only an essential characteristic of the physical fabric of resorts, but also an important element of their social and cultural construction (Bull and Hayler, 2009). Theatres and music-halls were built by ambitious individuals and groups keen to enhance their town's attractions, and their own fortunes (Walvin, 1978) and, in the 1930s, the additional craze of ballroom dancing had to be accommodated. The 1930s possibly represented the 'heyday' of the live entertainment phenomenon as the Second World War brought major difficulties for the resorts; many found it difficult to recover. In the immediate post-war period and the 1950s there were still enough tourists at the seaside to encourage investment in entertainment provision, but, by the 1960s, traditional forms began to be eclipsed by the public's desire to see the stars of the new mass medium – television – and, especially, pop stars (Williams, 1992, p. 11). By the early 1980s it was obvious that there were severe problems affecting the future provision of live entertainment at English seaside resorts, which were accused of promoting an outdated product (English Tourist Board, 1984). Hayler (1999) suggests that throughout much of the twentieth century the number of theatres in resorts has declined, but he has suggested that, while no longer a main reason for holidaying at the seaside, tourists still want to see the retention of theatres, no doubt because they see such facilities as part of the essential fabric of the traditional, authentic, resort environment (Hayler, 2004).

A range of other representations and symbols of the seaside also figure prominently in the popular imagination and perceptions of the authentic resort and objects of 'warm authentication' by tourists. This is worth noting because resorts are one of the places (theme parks also) where commodification is actively sought as a symbol of authenticity, rather than inauthenticity. Franklin (2003) refers to the 'objects and rituals of the seaside' that include ritual clothing, deck chairs, pony rides, souvenirs and postcards, in addition to the liminoid and ritual forms of behaviour, especially those associated with the beach, as discussed earlier. The picture postcard is a particularly useful cultural object which, although derided by some as an 'insubstantial piece of ephemera', can provide both a visual image of the physical environment of individual resorts and, through their humorous forms, representations of their use and users, as is evident in Wheeller's (2007) discussion of Eric McGill's saucy seaside postcards, 'A stick of rock, cock?'. As Gray (2006, pp. 84–5) points out 'postcards were particularly significant in establishing and developing the image of a place' and, in addition to the picture itself, the postcard also allowed, through the space for a hand-written message, holidaymakers 'to create their own representation of what was important in their own holiday experience, or at least what they wanted the reader to believe'. One of the distinctive aspects of many resorts was the Italian design influence; the ice-cream parlours and cafés with Gaggia machines producing 'frothy coffees' became emblematic of most seaside towns in the 1950s.

While this section has so far talked about cultural perspectives in general, the reality is that there exist some contrasting cultural representations. On the one hand, seaside resorts are seen to represent places of popular mass culture with their funfairs, amusement arcades, popular entertainments, bathing beauty contests and mass consumerism; places that some would argue display tastes that are 'common', 'vulgar' and 'low' (Urry, 1997; Agarwal, 1997). On the other hand, an alternative cultural perspective is that of viewing the seaside as a curative, restorative environment with its sea air, promenades and gardens, and vistas on which to gaze. Also, linked to this, different cultural interests include the artistic tradition associated with seaside resorts including, for example, the paintings of J.M.W. Turner (Margate has opened an art gallery – the Turner Contemporary – linked to his work), the works of Barbara Hepworth and the St Ives School.

The above paragraphs have described and examined the various physical and culturally constructed features that essentially characterise the seaside resort. Such features are recognisable and are deeply ingrained in people's perceptions; and it is on this basis that a case can be made for the existence of an authentic place. There are, however, a couple of important caveats to this proposition. First, this account is based on an extremely attractive experience of the seaside, one which is loaded with nostalgia and 'embroidered from… memories of childhood holidays' (Walton, 2000, p. 3). However, there are other, more negative, versions of the seaside experience, which according to Walton (ibid.):

> emphasise (for example) the loneliness and isolation of seaside retirement; the threat posed by erosion and flood to life, property and security in

> vulnerable areas; the rootlessness of the young seasonal worker in arcade, boarding-house or bar; the sexual fears and opportunities engendered by the transitory and complicit aspects of the resort environment, for residents and visitors alike; the earthy seediness of the less glamorous levels of the entertainment industry; the violence of gangs and protection rackets in places where a high throughput of visitors seeking services on or beyond the fringes of legality has established lucrative and tempting markets for the providers; the poverty amidst plenty endured by those for whom seaside employment is intermittent, unorganised, patronage-dependent and ill-paid; and more generally the conjunctions of defensive gentility, precarious respectability, chromium-plated glamour, earthy humour and raw sexuality which make up some of the distinctively seaside-flavoured cocktails of British popular culture.

The second qualification relates to the substantial economic and cultural change that resorts have experienced in the last few decades. British resorts continued to thrive until the mid-1970s; in fact, Demetriadi (1997) refers to the period from 1950–74 as the 'Golden Years'. However, following competition from foreign resorts – enabled by cheaper and simpler air travel, guaranteed good weather and superior accommodation – visitor numbers have declined, a trend further exacerbated by changes in values, beliefs and identities leading to British resorts appearing unfashionable and tasteless. Some of the entertainments that, at one time, were unique to resorts are now more widespread and thus seaside resorts have 'ceased to be distinctive as centres of spectacles' (Shaw and Williams, 1997, p. 4). Resorts moved from 'real to fake' (Lovell, 2013) as they became shabby and the funfairs and amusements began to seem anachronistic and out of place in the modern world. The final section will now examine the ways in which resorts have responded to decline and assess the extent to which they have refashioned themselves while still maintaining a sense of retreating authenticity.

Resort cycles, decline and regeneration

The issue, as seaside resorts change, is whether they still retain their sense of aura for tourists. As mentioned in the introduction and discussed at length in the chapter on gentrification, authenticity has been described as emergent (Cohen, 1988a), but, if this is the case, it can also retreat as places go out of fashion. As dynamic places, resorts have always undergone change, but over much of their existence such change was part of a positive trend, helping to establish them as distinctive leisure and entertainment centres, part of the process of creating the authentic resort. Butler (1980) first introduced a life cycle concept to tourist areas and it has subsequently been applied to seaside resorts and modified by others (Agarwal, 1994; Cooper 1992; Cooper and Jackson, 1989). The development of various resorts has been shown to mirror the form of the model. Wilson (1989, cited in Ryan, 1991), for example, has shown how the growth in visitor numbers to

Scarborough relates to six distinct historical periods between 1761 and 1988 that clearly relate to the Butler model. However, the model has been criticised for being too simplistic and limited for dealing with the complexity and diversity of resort development (Smith, 2004; Shaw and Williams, 2002). Rather than real to fake, some resorts could almost be said to move through cycles from real, to forgotten, to remembered, to nostalgic, to cultural destinations. Agarwal (1994, 2002) has been particularly critical; she argues that resorts may be subject to constant renewal so that their life is extended repeatedly because the coalitions of local interests are unable to accept the consequences of decline.

There is a unique character to a genteel or well-preserved resort such as Eastbourne and also a gritty authenticity (Zukin, 2010a), as discussed in Chapter 4, to those resorts which are run down, or less gentrified, such as areas of Hastings. Many resorts have experienced decline and could be viewed as being in the final stage of the life cycle, with the seaside tourist experience perceived as 'ageing and (lacking) competitiveness in the face of new products such as the Mediterranean package holiday' (Shaw and Williams, 1997, p. 9). Cooper (1997, p. 83) has identified a wide range of problems that have faced many British resorts, while Shaw and Williams (1997, p. 17) refer to a 'long slow decline in seaside tourism', with resorts being described as having a rundown appearance and looking tired and outdated with a poor destination image. Despite this view, others have argued that while British seaside resorts have been through a turbulent period of change, 'they are remarkably resilient, morphing into new lives for fashionable nostalgias and kitsches; homes for theme parks; new and desirable locations for weekend villas, condos and spaces of retirement and as centres for conferences and conventions' (Franklin, 2003, p. 137). Walton (2000) also points to their resilience and capacity for reinvention and, certainly in more recent years, there has been considerable emphasis on regeneration and renaissance.

The process of regeneration is not unique to seaside resorts but, unlike in many cities where tourism has been used as part of a strategy to regenerate local urban economies in response to de-industrialisation, in seaside resorts the situation is somewhat different. First, in resorts it is the tourism industry itself that has been under threat and, second, there is no suggestion that tourism is going to disappear or that the problem was ever as black as painted (Walton and Browne, 2010). It is also important to understand that resorts differ considerably in their nature and size and in terms of how they cope with decline and respond to the threat. The larger resorts, such as Blackpool and Brighton, for example, are of sufficient size to have the investment capacity to upgrade, and continually redefine and recreate their product (Shaw and Williams, 1997). Blackpool has regularly upgraded its attractions and its 42 acre Pleasure Beach Theme Park/Amusement Park is Britain's top visitor attraction; in the last 20 years it has seen several multi-million pound new developments introduced, including the £12 million Pepsi Max Big One, one of the tallest and fastest rollercoasters in Europe, and the £15 million Valhalla, described as the 'biggest most spectacular dark ride ever to be constructed by mankind'. In contrast, investment in some medium-sized and small resorts is much lower and halting decline is more challenging. However, several small resorts have developed successful niche markets – such as surfing in Newquay and Whitby capitalising on its literary association with Bram Stoker's Dracula and the associated Goth festivals. While these at first seemed staged, they are now central to the character of the place.

Much regeneration policy is driven by economic and social imperatives – especially job creation, with some coastal resorts having been identified as having the worst levels of multiple deprivation in Britain, being on a par with deprived inner-city areas. In fact, 26 of the 37 principal seaside towns in England have an overall level of deprivation greater than the English average (Department for Communities and Local Government, 2008). Along with various inner-city areas, many coastal resorts also have relatively high levels of multiple deprivation and, as with cities, much regeneration is aimed at reducing this, as outlined in publications such as *Tomorrow's Tourism* (Department of Culture, Media and Sport, 1999), the Coastal Communities Alliance volume edited by Walton and Browne (2010) and the Department for Communities and Local Government publication, *Strategy for Seaside Success: Securing the Future of Seaside Economies* (2010), as well as the earlier *Sea Change* (Department of Culture, Media and Sport, 2008). These policies place the sense of authentic seaside place at the heart of the strategy.

While much attention has focused on economic and social issues, there is also a cultural dimension; Dryburgh (2010) states that initiatives to commission art and cultural projects and programmes have escalated in recent times, with cultural activity operating in a variety of ways (Evans and Shaw, 2004; Kennell, 2011; Walton and Browne, 2010). In fact, the government's *Sea Change* programme (Department of Culture, Media and Sport, 2008) has been supporting wider economic regeneration in a number of coastal areas through investment in culture and heritage. *Sea Change* committed the Department of Culture, Media and Sport to fund projects aimed at placing culture at the heart of regenerating England's seaside resorts by investing in arts, public space, cultural assets and heritage projects. Similarly, the Heritage Lottery Fund continues to support a wide variety of heritage schemes in coastal towns, ranging from parks and regeneration, to museums and community projects and, since 1997, has given over £234 million to 864 projects in English coastal resorts to support their regeneration, including over £100 million to coastal resorts in deprived areas including Blackpool, Falmouth, Great Yarmouth, Hastings, North Shields, Penzance, Redcar, Saltburn-by-the Sea and Southport (Department of Culture, Media and Sport, 2010). Cultural placemaking was discussed in Chapter 5, but it is worth mentioning English Heritage's policy *Shifting Sands* (CABE, 2003). This document set the bar for architectural design at the beach, aspiring to preserve iconic buildings such as the art deco De La Warr Pavilion in Bexhill-on-Sea. The report also featured new fusions of modern architectural styles, such as the Beach Hut Café and Extreme Academy at Fistral Beach, using colour-stained wood resembling driftwood, with sympathetic designs suggestive of boat hull and surf board, referencing the authentic character of place.

Other developments in recent years, however, have reduced the distinctiveness of seaside resorts as they have become important providers of accommodation for retirees, commuters, social security claimants, asylum seekers and artists. Seaside retirement developed particularly rapidly in the 1960s; by 1971 35 of the larger resorts had more than 20 per cent of their population above pensionable age (the national figure being 16.1 per cent), with several resorts having percentage figures at more than twice the national norm, Bexhill being as high as 44.2 per cent (Walton, 2000, p. 158). As Walton (ibid., p. 46) points out,

> the resort elements in local economies were shrinking even as the towns continued to grow. At the end of their product life cycle as resorts, they were transmuting into other kinds of place, dormitories and retirement centres which happened to have seaside locations.

Davidson (2015) states that 'there are now seaside towns across the UK which can fairly be called part of the commuter belt – and where half the people you bump into on the seafront have a day job in the heart of the city'. For example, people can now reach London by rail in under one hour from many of the seaside resorts in Kent, Sussex and Essex; mobile devices, which allow people to work during their journey, further reduce the inconvenience of travelling. Such developments may well be part of a broader shift which is seeing seaside resorts becoming less 'tourism centric', as advocated by Agarwal and Brunt (2006) – being transformed into 'towns by the sea' instead of resorts.

Conclusion

In seeking to establish whether resorts might be regarded as authentic places, it is their special character that lies at the heart of such a claim and whether the key ingredients of this physical and cultural environment are still evident. The seaside resort is perhaps the quintessential tourism place because of its unique set of environmental characteristics, cultural construction and perceived importance for leisure. For over 200 years, coastal sites have been attracting visitors and for many people the seaside resort was, and continues to be, synonymous with holidays and day trips.

This chapter has described the attributes of resorts which make them special, as well as examining the various factors and processes that have combined to produce them. While the attractions and fortunes of resorts have declined in recent times and initiatives to offset such decline have, in some instances, created additional functions and modified the typical resort morphology, many of the features of the classic seaside resort remain. Beaches, promenades, ornamental gardens, funfairs, amusement arcades, theatres, fish and chip shops, seafood stalls (Figure 6.2), shops and kiosks selling sticks of rock, candy floss and various seaside memorabilia… are still prominent and continue to evoke nostalgia and attract substantial numbers of visitors.

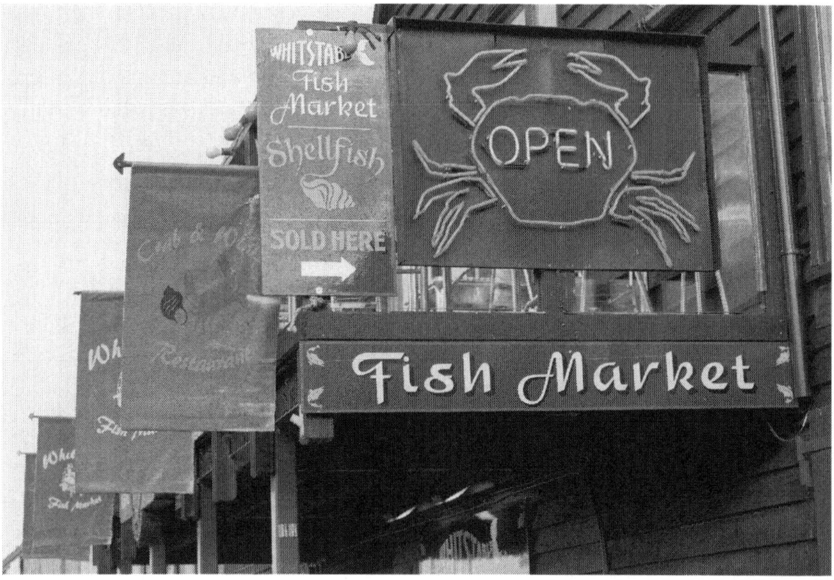

Figure 6.2 Whitstable oysters
Source: Canterbury City Council

In addition, resorts are a surprising fusion of the exotic, Edwardian grandeur, working-class pop culture, gentility, contemporary art and design icons, which make them key destinations for hipsters and key targets for gentrification. Although the number of long-stay holidaymakers has declined in the last few decades, there is still a sizable number of people who continue to enjoy the traditional holiday by the seaside and it remains an important destination for day trips; Blackpool Pleasure Beach, for example, is still Britain's most popular attraction. It could be said that that part of the uniqueness has faded as other places have developed attractions that can compete with the seaside. However, as Franklin (2003, p. 137) points out, although the seaside has become:

> one of very many alternative types of pleasure.... it remains one of the most important, precisely because it retained its basic ritualistic formula: it maintained its liminoid space of the beach, its ritual devotion to the sea, the sun and the body and its unhurried basic structure as a ritual of passage. When people say they need to get away, when people conjure an ideal holiday or break, in their imaginations seaside is still a dominant evocation.

Resorts are far from static and throughout their history they have experienced significant physical, cultural and economic change, a process akin to 'riding the big dipper' according to Williams and Shaw (1997). Nostalgia is a powerful force, it suggests that people can view places selectively, transposing what they want to see onto reality. Although resorts may change, if they retain the light, sea, piers and promenades, there are many prompts for 'warm authentication' which enable them to retain their sense of more-than-human aura and reinforce the seaside idyll.

7 The rural idyll

From cultural countryside to constructed authenticity

The rural environment is the result of physical and economic processes, especially agriculture, but its attractiveness is also due in large measure to its social and cultural construction (Cosgrove, 1984; Bunce, 1994; Urry, 1995a; Aitchison *et al.*, 2000). To many people the countryside has a special appeal involving an image that is often described as idyllic (Newby, 1980; Mingay, 1989). The construct of the English rural idyll remains a powerful influence on people's perceptions of the rural environment, responsible, in large measure, for substantial urban–rural migration and for attracting millions of visitors to the British countryside every year, as well as having profound effects on the way the environment is managed. The chapter discusses the history of the idyll, which has been reinforced through art and literature, and more recently through television and advertising – portraying the rural environment as superior in nearly every respect to that of the town and city (Williams, 1973; Wiener, 1981). The chapter illustrates the 'emergent authenticity' (Cohen, 1988a; Bruner, 1994) of the rural idyll in art and literature and the influence of events such as the First World War, which cemented the nostalgic idea of an idealised homeland. The chapter also explores concepts of manicured land created by designers such as Capability Brown and their impact on landscaping and the development of parks and gardens, as well as methods of modern farming which are shaping the actual rural environment.

The rural idyll is a powerful engine of idealisation, with popular mass appeal between perception and reality. While the chapter focuses on the British idyll, its essence has international resonance, as demonstrated by syndication success of the television programme *The Great British Bake Off*, which celebrates representations of a village aesthetic such as bunting and gingham. The destruction of the rural idyll was most eloquently demonstrated in the 2012 Olympic Games Opening Ceremony directed by Danny Boyle, in the sequence 'Green and Pleasant Land'. Rural scenes were performed, complete with floating clouds, cricket matches, men in shepherd's smocks and a small rural hamlet. They were replaced in the 'Pandemonium' sequence by the Victorian industrial revolution's machinery, iron casting, a beam engine, omnibus and factory towers which rose from a scarred earth, destroying the countryside with an intentionally violent effect. 'Pandemonium' clearly showed the fault line and trauma of industrialisation, which still runs through the national psyche, creating a non-specific sense of loss and a nostalgia which percolates into the tourist search for authenticity.

However, as various authors have highlighted, the idyllic image is at odds with reality (Newby, 1980; Thrift, 1989); and this chapter focuses on the rural idyll as an example of Wang's (1999) 'constructed authenticity', revealing that the actual countryside is not always as pleasant, attractive and welcoming as the image would suggest. The discussion of the rural idyll acts as contrast to the modernist illusion of the urban, modernist phantasmagoria discussed in Chapter 5. Instead, the rural idyll evokes a gentler imagined place, looking back as opposed to forward, with a similar sentiment to the celebration of old-fashioned pastimes discussed in Chapter 4. Here it was contended that the more that globalisation accelerates, the more likely some people are to slow and authenticate life through adoption of pre-industrial practices. Selwyn (1996) discusses hot and cool authenticity and Cohen and Cohen (2012) argue for hot and cool authentication; this chapter focuses on a 'warm authentication' based on a nostalgic gaze which creates and perpetuates an imaginary place. Relph (1976 [2008]) argued that nostalgically constructed places are essentially inauthentic and the chapter explores the rural idyll as an example of a 'fake-real' notion (Fjellman, 1992; Lovell, 2013) of a reality based on nostalgia, highlighting the various ways in which the reality differs from the idyllic perception. The chapter complements the discussion of historic cities, demonstrating different ways in which an aura was constructed, leading to a notion that there is a diffuse 'original' countryside, engendering a search for authenticity in those pursuing the idyll. While the chapter focuses on the British idyll, the concept of a pre-industrial, simpler era and place where the past is preserved also has wider relevance.

The creation of the idyll

The idyllic image of the British, and especially English, countryside incorporates picturesque, well-ordered landscapes, healthy lifestyles, close-knit friendly communities, tranquillity and stability, the latter idea being particularly important in a world that, in reality, is dominated by rapid change. Two other ideas also figure prominently in this popular image of rural ways: simplicity and involvement with nature. Rural dwellers are often perceived as leading simple, some might say even 'primitive', lives, as celebrated in novels such as Flora Robson's (1943) *Lark Rise to Candleford* and Elizabeth Gaskell's 1853 novel *Cranford*, where village life is free from the complexities of urban living, with its pollution, congestion, crime and financial crises and where modernity, in the form of the railway, intrudes rudely. Farming, the principal activity of the countryside and one that has been practised for thousands of years, is concerned with plants and animals, with soil, climate and the changing seasons and, thus, no one is thought to be in closer contact with nature than the farmer and farm worker.

As various authors in Mingay's (1989) edited text point out, perceptions of countryside change with time, authenticity emerging as each generation of country dwellers and observers value land differently. Nevertheless, there remains an enduring, popular image of a rural landscape that is somehow more natural, more peaceful, more spiritually satisfying than the urban. In the poet William Cowper's words: 'God made the country, man made the town'. Many suggest that this image can be traced back to the Romantic movement of the early nineteenth century and, as can be seen below, such art and literature certainly did much to strengthen the perception. However, the image is of much longer standing. Williams (1973), for example, traces the way in which this powerful Arcadian view of countryside has been portrayed in literature throughout the ages, for example through the Romantic movement, but also as far back as the ancient Greeks. There is also a counter view that the rural environment is boring and backward, but the more popular image is the idyllic one which, in addition to attracting large numbers of tourists to the countryside, has also persuaded many people to move from towns and cities to live there (commuters, retired people and second home owners) in search of the 'good life'.

The explanation for the emergence of this widely held image may involve several factors including art and literature, the influence of key image-makers and 'educators', and the various psychological needs of people. As mentioned above, throughout the centuries literature has always painted an attractive view of the countryside, warmly authenticating it using tropes and motifs. In the past those who wrote about the countryside either belonged to the upper classes, who derived their wealth from country estates, or were patronised by them, for, as Williams (ibid., p. 33) points out, 'poets have often lent their tongues to princes who are able to pay or to reply'. In either event, writers and poets were, therefore, most unlikely to write unattractive accounts of such places and the communities that lived there and, thus, critical, negative writings of countryside are relatively rare. Furthermore, as Keith (1989, p. 77) suggests:

> A country's literature does not necessarily reflect the historical actualities of the time in which it is written, since writers are influenced by conventions and the traditional practices and expectations of their art as well as by what they see around them.

Culturally constructing the idyll

A tradition of pastoral poetry and writing was established which waxed lyrical about the virtues and attractiveness of the rural environment and its practices, implicitly suggesting that it is more authentic than the industrialised urban lifestyles of the time. Many of the nation's most well-known poems (and lines) by such poets as Brooke, Browning, Clare, Housman, Keats, Shakespeare, Wordsworth…, which are ingrained in people's education and culture, involve idyllic notions of countryside. Wordsworth's *The Daffodils*, for example, is usually included near the top of surveys of the nation's favourite poems, as are Keats's *To Autumn* and *Ode to a Nightingale*. Various writers (Strong, 2012; Mount, 2013; Aitchison *et al*., 2000) highlight the significance of the Romantic poets in strengthening the idyllic vision, but the longing for the pre-industrial age continued throughout the nineteenth century with the great Victorian poets (e.g. Tennyson, Swinburne and Browning) and also in what Strong (2012, p. 153) refers to as 'the golden age of the novel' where 'writers too invented a rural England and peopled it with a veracity that still haunts our imagination'. Other publications were also influential and Strong (ibid., p. 156) cites the importance of *Country Life*, first published in 1897, where 'week in and week out, city-dwellers who aspired to life in the country could feast on images of historic manor houses, gardens and villages, and learn about rural sports and customs'. As he emphasises, 'the fact that *Country Life* is still in existence testifies to the longevity of the rural idyll and its continuing appeal to those living in cities'.

Some authors have also romanticised the people and community. Elizabeth Gaskell's novel *Cranford*, modelled on Knutsford in Cheshire, features a predominantly female village full of characters and their intrigues. Hardy's novels such as *The Mayor of Casterbridge* incorporated vernacular country speech patterns. While there were many literary caricatures and country life was also depicted as difficult, there was an element of nostalgic authenticity about the depictions in the speech, wisdom, customs, relationship with nature and folk remedies employed by characters such as Victoria May 'Queenie' Warrener Turrill in Flora Robson's *Lark Rise to Candleford.* The contrast between the rustic, harder, simpler life village of Lark Rise and the precision, pride and comparative modernity of the small town of Candleford depict people torn between the two ways of living.

While the nineteenth century produced a wealth of literature emphasising the rural idyll, the tradition of idyllic writing did not end there but continued throughout the next century and to the present day. A wealth of coffee-table books feature attractive images of various aspects of rural life, many of them produced by the various organisations that exist to campaign to protect the rural environment such as the National Trust (NT) and Royal Society for the Protection of Birds (RSPB), as discussed later in this chapter. Their magazines and those of the Ramblers' Association and Royal Horticultural Society promote interest in plant life, gardening, garden tourism and exhibitions, such as Tatton Park and Chelsea, reinforcing the concept of a glossy, beautifully photographed outdoors. A vivid example that epitomises the genre can be seen in two books concerning poetry associated with the English countryside, *Poems of the Countryside* (1987) and a sequel, *Green and Pleasant Land: Poetry of the English Countryside* (1989). The format of the books, each 125 pages long, involves pages of poetry with a rural theme, interspersed with landscape paintings and drawings of idyllic rural scenes or features by Gordon Beningfield. Almost 200 poems are included across the two books, drawn from the work of poets spanning four centuries, with virtually all expressing an idyllic perspective.

In addition to poetry, there has also been a strong tradition of children's publications that has reinforced the idyll, warmly authenticating the countryside. Nostalgia is based on the idea of reflection, gazing back selectively to view eras as idyllic and notions introduced in childhood can retain a powerful influence. Many children would have been familiar with such classics as Kenneth Grahame's

Wind in the Willows, A.A. Milne's *Winnie the Pooh* and Arthur Ransome's *Swallows and Amazons*, as well as Beatrix Potter's various stories, with their anthropomorphic wildlife, picnics and interesting adventures in a safe and benign environment. As Drabble (1979, p. 247) has noted 'children's books are strikingly rich in evocations of idyllic landscape' and even children who live in towns 'can travel to the country in works of the imagination'. More recent examples of this type of children's literature would include Jill Barklem's richly illustrated *Brambley Hedge* stories. There also exists a wide collection of children's nature books, such as those published by Usborne and Ladybird, which have helped to instil a love of nature and the countryside among generations of children and young people. The Ladybird *What to Look For* seasonal series – illustrated by Charles Tunnicliffe, the famous wildlife artist – is a classic example. Tunnicliffe's illustrations, for example, are both colourful and idyllic and almost certainly would have left a powerful impression on young readers. Children's television programmes, such as *Postman Pat* and *Candlewick Green*, would no doubt further encourage young minds to assume that the countryside was idyllic, in this case inhabited by friendly, harmonious communities.

During the last couple of centuries there has also emerged a strong tradition of landscape painting which has further reinforced the idyllic image of the countryside. The paintings of John Constable are often cited as the epitome of this tradition, a style created in the second half of the eighteenth century, often referred to as the Picturesque movement, 'in which the elements of landscape played a major rather than a subordinate part, working on canvas what the landscape gardeners were working on the ground' (Treble, 1989, p. 50). Such paintings often involved picturesque landscapes but also 'scenes of cottage life, featuring rustic simplicity, sometimes in picturesque poverty, sometimes in virtuous and provident though rude comfort' (ibid., p. 53). Baldwin *et al.* (2004, p. 149) describe Constable's famous painting *The Hay-Wain* as presenting 'an undeniably positive view of the countryside' where 'the rural is represented in terms of harmony, beauty, stability, tradition, peace, innocence and virtue'. Mount (2013, p. 145), in his examination of the Picturesque movement, considers it to have contributed enormously to 'a peculiarly English type and sense of beauty', and that it continues to do so. He suggests that its specific nature 'encompasses an apparently happenstance, wild, accidental beauty, even if the beauty is choreographed by a hidden human hand'.

In her account of the 'Victorian picture of the country', Treble (1989, p. 53) lists Victorian artists and specific paintings whose 'prettily sentimental view of country life was almost as mythical to its contemporaries' as it is to today's observers. Nevertheless, irrespective of their veracity or rather because of the 'real-fake' utopias they promoted, such paintings caught the popular imagination. She cites the example of William Collins (1788–1847), who received early fame with his painting *The Disposal of a Favourite Lamb* (1813), which was engraved twice that year and some 15,000 of the smaller prints alone were sold. Collins 'capitalised vigorously on the fashion' with subjects like *Rustic Civility* (1832) and *Cottage Hospitality* (1834); one of his most popular pictures was *Happy as a King*, depicting 'apple-cheeked children [scrambling] on a five barred-gate, clean, dimpled, and robust despite their rags'. Treble (ibid.) suggests that such 'innocent and healthy examples of the delights of rural poverty' provided a clear comparison 'with the all too evident horrors of urban squalor which presented themselves on every side to Collins's largely urban audience'. Another key artist cited by Treble is Myles Birket Foster (1825–1899) whose picture *Children Playing* involved 'carefree, pinafored children [dashing] down a rolling hillside, pet dog scampering beside them, towards the village pond'. Foster's watercolours, which he continued to paint for a quarter of a century, after 1859 usually involved 'children gambolling in nature's garden', gathering flowers or berries, playing on stiles or old logs or in front of cottages, listening to stories, shelling peas or tending chickens' and they 'imprinted on the English consciousness an idyllic fantasy of rural existence that was only partly countered by the social realists of the 1870s' (ibid., pp. 53–4).

Also linked to a shared history are feelings of nostalgia for an earlier, gentler, pre-industrial age. In the late nineteenth century there developed a strong revulsion to the excesses of urban industrialisation and 'the countryside was seen to embody the timeless essence of "Olde England", encapsulated in the ancient manor house with its attendant village and parish church' (Strong, 2012, p. 150). Strong (2012) also highlights the role of contemporary photographers in portraying this vision and cites the example of Henry Peach Robinson (1830–1901), who portrayed rural life as 'seemingly unchanged: an England peopled by farmers, craftsmen and labourers pursuing their cyclical toil through the seasons' (ibid., p. 150). Thus the concept of a 'golden age' was established when a contented pastoral way of life was possible for most people, with village communities living in social harmony and interdependence. In part, it is a recollection of childhood, when summers seemed brighter and anxieties fewer, but also it is the 'magical evocation of a land that needs no farming' (Williams, 1973, p. 28).

The gaze seems to be locked into an Edwardian, pre-First World War view of the countryside which emerges periodically, rediscovering and reauthenticating rural heritage practices and styles anew. For example, during the late 1960s and early 1970s books such as *The Country Diary of an Edwardian Lady*, by Edith Holden, capitalised on this 'retrochic' (Samuel, 1994) which was also epitomised in the floral prints, sprigged muslins and Edwardian styles of Laura Ashley. Different styles fused into a reworked Olde England, as demonstrated in Fairport Convention's 1969 album *Leige and Lief* and Mike Batt's reworking of old folk tales in Steeleye Span's 1975 album *All Around My Hat*. The adoption of the rural way of life was revisited in the 1970s television programme *The Good Life*, where Tom and Barbara became self-sufficient in the suburbs in a search for a more authentic way of living than working in advertising. *The Good Life* interestingly pre-dates the hipsterisation of places through anachronistic methods of production, which is discussed in Chapter 4. Periodic movie remakes of Thomas Hardy's novels such as *Far From The Madding Crowd*, while incorporating some notes of hardship, indicate the enduring appeal and popularity of the idyll. Furthermore, the longing for an earlier, rustic way of life is perpetuated through performance, with customs such as May Day, or the performative authenticity of Morris dancing following traditional patterns and choreography.

Allied to the Picturesque movement in art was 'the revival of English vernacular architecture, which began in the 1870s and drew on the buildings of Tudor England and those of the age of Queen Anne which were bracketed together as Old English' (Strong, 2012, p. 183). This style replicated old methods and materials, using used stone, brick and half-timbering, massive chimneys, great halls, inglenook fireplaces and complex enveloping roofs. In bringing a number of different periods and strata together, it created an 'eclectic view of Old England' which, according to Strong (2012) ennobled the black and white timber-frame house as much as it beautified the humble country cottage, seen to enshrine the domestic virtues of an earlier, more gentle, pre-industrial age. Such perceptions were further reinforced by Helen Allingham's romantic watercolours of cottages and their pretty gardens in her *Cottage Homes of England* (1909). It should also not be forgotten that the half-timbered, Tudorbethan style became a popular form of aspirational middle-class housing development in the suburbs in the 1920s and 1930s, seeking to emulate a rural retreat that was, for many, not yet possible at that time. It is impossible not to recall Dame Edna Everage's description of a Morris Minor Traveller, which featured a timber frame, as a 'Tudorbethan car'.

Advertisers have also used rural imagery to help sell their goods, ranging from chocolate bars to motor vehicles and, in so doing, have continued to reinforce the idyllic image. Roscoe (1996, p. 89), for example, highlights an interesting example of this in the way oil companies, in the period 1930–70, appropriated 'representations of Britain and its landscapes to market their products'. Car ownership increased substantially during the 1950s and 1960s and this, together with the expansion of the motorway network and improvement of other roads, led to mass motor touring. As Roscoe (ibid., p. 90) points out 'by the mid-1960s, motor touring and the British countryside had become inextricably linked in the minds of most people' and 'coinciding with this growth in motor touring was a distinct flourishing of natural history and countryside-based advertising (in the form of posters, books and television programmes) sponsored by oil companies'. The oil companies realised that:

> if motorists could be shown attractive representations of the countryside and its natural history, then they were likely to wish to observe them at first hand. The more they were encouraged to tour in this fashion, naturally the greater the volume of petrol sold.
>
> Ibid.

Shell-Mex and BP published a wide selection of books about the countryside during the 1950s and 1960s, perhaps the most famous and influential being the *Shell County Guides*. These, together with *The Shell Guide to Britain* and individual country *Guides* (England, Scotland and Wales), were sponsored by the oil company Shell and, as Heathcote (2011, p. 4) points out, 'their unique quality comes from the combination of literature, antiquarianism, art and design with a devotion to expressing and revealing the pleasures of driving around counties by minor roads', with a vision involving 'a sense of holiday'. According to Heathcote 'it was the special gift of the Shell Guides to represent as an everyday fact the elegiac oddness of the British landscape and to imply that this is part of the national identity' and they were clearly designed to appeal to the traveller, as 'a seeker of the authentic place' (ibid.).

National identity

National identity and concepts of Englishness are certainly key ingredients that fuel the continuation of the idyll (see e.g. Matless, 1990; Paxman, 1998; and Strong, 2012). Paxman (1998, p. 142), for example, reminds us of John Major's famous 1993 speech, with its references to a country 'of long shadows on county grounds [and] warm beer' evoking visions of the 'national game' which Strong (2012, p. 187) reminds us is 'emblematic of England and its pre-industrial rural past with its immemorial tableau of the village green, church, manor house and cricket pavilion and its cast of squire, parson, gentlemen players and villagers'. Paxman (1998, p. 142) and Strong (2012, p. 163) also cite a speech made 70 years earlier by Stanley Baldwin who refers to:

> The sounds of England, the tinkle of the hammer on the anvil in the county smithy, the corncrake on a dewy morning, the sound of the scythe against the whetstone, and the sight of a plough team coming over the brow of a hill.... The wild anemones in the woods in April, the last load of hay being drawn down a lane as the twilight comes on... and above all, most subtle, most penetrating and most moving, the smell of wood smoke coming up in an autumn evening... These things strike down into the very depths of our nature... These are the things that make England.

Both speeches were aimed at evoking national sentiments based on a picture of England rooted in eternal rural imagery.

A further example that Paxman (1998) and others (Matless, 1990; Strong, 2012) highlight is how British soldiers in both world wars were subtly led to believe that it was the countryside (the land of churches, fields, gardens and villages) that they were fighting for. Soldiers were often provided with books of poetry with many of the poems portraying idyllic notions of countryside. Paxman (1998, p. 146) talks of Rupert Brooke's famous First World War poem involving a marriage of some of the strongest ideas of Englishness – goodness, home and countryside – encapsulating 'exactly how the English wanted to see themselves' and he goes on to describe how other writers 'kept alive the idea that the soul of England lay in the countryside' (ibid., p. 147) (e.g. Housman's *A Shropshire Lad* and Massingham's many books telling the English people that the industrial revolution had 'destroyed "the true England"', while the countryside was 'a source of our daily bread and the indispensable foundation of our national well-being'). Strong (2012, p. 172) suggests that 'the ruralist vision that we find in Brooke and the other war poets could be put to the service of national and imperial goals in times of war', and that, in the Second World War, 'the rural vision of England once again became the touchstone of patriotism'. Furthermore, popular wartime songs reinforced this message, such as the following famous lines that Strong (2012, p. 146) highlights:

> There'll always be an England
> While there's a country lane
> Wherever there's a cottage small
> Beside a field of grain.

These notions of the rural, linked to concepts of Englishness, are also part of the nation's history and national pride, conjuring up visions of the longbow men at Agincourt and the English oak of England's naval ships; and Strong (2012, p. 186) highlights the role of history teaching in strengthening this common cultural inheritance.

Finally, it has also been suggested that many people are happy to maintain such images because they act as a means of escapism, as a counter to the seemingly harsher reality of urban living (Newby, 1980; Treble, 1989). As Newby (1980, p. 14) has pointed out: 'Somewhere, it is believed, at the far end of the M4 or the A12, there are "real" country folk, living in the midst of "real" English countryside in – that most elusive of all rustic utopias – "real communities"'. Such ideas have thus become ingrained in people's consciousness; they are continually being reinforced by the publicity of tourist organisations and conservation/preservation bodies such as the NT, RSPB and the Campaign to Protect Rural England and in programmes with vast popular appeal such as *Countryfile*.

The previous sections have shown how different components of the rural idyll were constructed and became self-reinforcing. However, as Thrift (1989, p. 26) argues, this bucolic vision of the English countryside is a 'fundamentally constructed one, comprised of elements that never existed together historically' and, as Newby (1980, p. 14) suggests, is 'located only in the minds of those engaged in the search for it, on a few calendars and chocolate-box lids – and in the wholly misleading paintings of John Constable'. In fact, all the key elements of the image – the picturesque landscapes, the 'natural' quality, the unchanging tranquillity and close-knit friendly communities – are fake-real, as the following sections will show.

The modern farmed landscape

While the idyllic vision of the countryside is of a picturesque farmed landscape that has remained unchanged for centuries, the reality is very different. For many centuries agriculture has shaped the rural landscape but, despite popular perceptions that the countryside is unchanging, such landscapes have not remained static. For example, in the late eighteenth and early nineteenth centuries, in many areas of England, open field systems were replaced with enclosed fields, a period of change that, at the time, evoked considerable criticism concerning its effect on the landscape (see e.g. Cobbett, [1830] 1967). The resulting farms containing both pasture and crops in relatively small fields often acquired a 'patchwork quilt' appearance, with an abundance of hedgerows and associated wildlife. These landscapes remained largely unchanged for the next 150 years and, just as the open field system beforehand, came to be regarded as 'traditional'; it is this varied, mixed farming landscape that has constituted the visual attraction in people's perceptions of countryside. However, in more recent times another set of changes has occurred that has dramatically affected agriculture and related landscapes.

To guarantee sufficient supplies of key food stuffs at reasonable prices, following the Second World War, government introduced a range of policies to support farmers such as generous guaranteed prices and grants to encourage 'desirable' practices. Although they achieved their immediate objectives of making farms more efficient and increasing food production, they also produced substantial landscape change and wildlife destruction. From the late 1940s agriculture experienced what some have referred to as the 'second agricultural revolution' – what others have described as 'productivist restructuring', involving the intensification of farming, increased scale of operation and increased specialisation (Ilbery and Bowler, 1998; Ilbery and Maye, 2010), with the enlargement of farms and fields having a substantial influence on the nature of farming landscapes due to the removal of thousands of kilometres of hedgerows. Furthermore, the destruction of small woodlands, draining of wetlands and the 'improvement' of heathland to extend the arable area also had a profound impact on landscapes in certain areas. In addition, improved machinery and agro-chemicals have allowed farms to specialise in those enterprises for which they possess a comparative advantage; thus, farms and associated landscapes have become more uniform and key habitats and associated wildlife have become degraded. Rather than having a patchwork quilt appearance, in some areas the modern landscape has been likened to that of a 'prairie'. Modern farming has become more a rationally calculated business than an 'Arcadian way of life'.

A different village

Another key aspect of the rural idyll is the image of the ideal community, incorporating a sense of neighbourliness and friendship and a sense of belonging, with village life proceeding at its own pace, with its own priorities, undisturbed by the problems of the outside world and with the 'timbered cottage and quaint country pub, clustered around the ancient church and village green, summarizing our rural heritage' (Newby, 1988, p. 31). As Newby further points out, 'as a result most of us feel that the rural village is a real community'. In his earlier text, *Green and Pleasant Land* (1980, p. 154), Newby explores the concept of 'community', suggesting that it can mean different things to different people. Village community can signify nothing more than a type of settlement, but it can also refer to the pattern of social relationships which exist within this locality – for example, everyone knowing everyone else. However, this does not describe the nature of such relationships, such as the level of harmony or whether people share similar values. It is these aspects that are encompassed by Newby's suggested third meaning of 'village community', involving 'a disembodied "spirit" of community, a sense of belonging, of sharing a social identity in a spirit of friendliness and common emotional experience'. It is this third meaning, which he suggests would be more aptly referred to as 'communion', that is involved when people talk about the 'village community' and decry its decline.

Although, in the past, when the village was almost entirely an agricultural community, it is likely that social relationships were close-knit, this could well have been due more to need than desire, as a result of having to cooperate and help each other in order to survive: what Williams (1973, p, 104) refers to as a 'mutuality of the oppressed'. It is certainly the case that village life for most inhabitants was far from attractive, with gross poverty the norm and hunger commonplace; producing food did not necessarily increase a person's chances of consuming it. Evidence of these harsh conditions can be seen in the numerous rural riots (e.g. the Captain Swing riots of 1830) and in certain contemporary writings (see, for example, some of the descriptions of the early nineteenth century in William Cobbett's *Rural Rides* and of the early twentieth century in Ronald Blyth's [1972] *Akenfield*, where many 'horsemen, ploughmen and field workers… fled the wretchedness of the farms in 1914' to join the army [ibid., p. 33]). When various artists and poets were extoling the outward attractiveness of the country cottage, Robertson Scott (1925, p. 21), for example, was observing that 'much of the misery of the hamlet is hidden from the passer by', with cottages being invariably cramped, cold and damp, with outside privies and virtually no other conveniences that most people today would regard as essential. Even in more recent times, agricultural labourers remained one of the poorest-paid workers and, through their relatively remote locations, suffered a host of accessibility problems as well (Moseley, 1979).

However, whatever village life was like in earlier times, today it is certainly very different from the perceived image. The 'occupational community', based largely on the single activity of farming that is a key component of the idyll no longer exists, and has not done so for some time (see Newby, 1980; Rowley, 1994). Villages have been experiencing change since the industrial revolution, but after the Second World War a period of particularly rapid change left them totally transformed. Many farming jobs have disappeared due to the industrialisation of agriculture and as traditional rural dwellers sought new forms of employment elsewhere, their numbers were replaced by newcomers (commuters, retirees and second home owners) lured to the countryside by the perceived attractiveness of village life. This movement and its effects on the nature of the village has been discussed by various sociologists and geographers (Pahl, 1965; Newby, 1980; Clout, 1981) and it has been, and remains, 'a potent force underpinning the social reconfiguration of rural Britain' (Gallent *et al.*, 2015, p. 135). The newcomers had no connection with the land and, as well as creating new mixed communities, their middle-class values and lifestyles and high level of emotional involvement with the 'rural idyll' (Woods, 2005) often brought conflict, centred on a clash of cultures and divided opinion as to the future needs and direction of many villages (Gallent *et al.*, 2015, p. 137). In more recent times, the counter-urbanisation process has gathered even further momentum and, rather than simply countering rural emigration, in-migration is producing a net growth of population in rural areas. Between 1981 and 2003 the population of rural England grew by 14.4 per cent, involving an average annual net migration of 30,000 people into rural areas. However, according to Best and Shucksmith (2006), these net flows may have become even larger since 2001; in the year 2013/14 net internal inward migration to predominantly rural areas in England was more than 60,000 (DEFRA, 2015). The social transformation of the village is continuing apace; an analysis by Phillips and Smith (2015) of rural social change using census data from 2001 and 2011 suggests the middle classes are continuing to colonise large swathes of rural England, with about 70 per cent of rural areas now having more than half their residents employed in middle-class occupations. This rural gentrification is reaching deeper into the English and Welsh countryside than ever before, but Phillips and Smith also identify what they describe as belts of 'super-gentrification' that have turned villages on the edge of London and other major cities into 'hot spots of increasing affluence'. The original working-class populations have thus become relatively small and what remain are becoming increasingly marginalised, with any notion of a close-knit community receding even further into the distant past. In addition, the continuing loss of the pub, post office, church and other amenities typified in the idyll are further eroding the village as a community.

One of the other enduring aspects of the rural idyll, and one that figured prominently in some of the artistic depictions that were described earlier in this chapter, is the attractive environment the countryside and village life afforded children. As Matthews *et al.* (2000, p. 151) point out, the idyll 'resonates with the lives of young people in general. Characterised by mutual support, harmony and integration, the idyll suggests a fulfilling and satisfying childhood for all'. However, Valentine (1997), for example, in her study of the opportunities for children in Wheldale, showed that parents contradicted popular constructions of the rural as a safe, harmonious place and contested assumptions about the idyllic nature of rural childhood.

A manicured countryside

Much of the British countryside is a farmed, working landscape which, as already explained, is an artificial environment, a landscape created for food production from its original natural, woodland state. Nevertheless, despite its artificiality, the agricultural environment is generally regarded as authentic countryside. The clearing of woodland for farming began in Neolithic times and thus the idea that a farmed landscape is authentic is of long standing, even though its precise form has changed at various points in history. There are, however, some areas of countryside that have been specifically created for other purposes but which still figure prominently in people's views of the rural. These include parks and gardens, often part of the country estates of the landed gentry and designed for their pleasure, and country parks, established following the 1968 Countryside Act and specifically planned to provide for a variety of outdoor leisure activities in a countryside setting, primarily in areas at reasonable distance from urban areas (Glyptis, 1991; Harrison, 1991). Both were designed to give a semblance of 'natural countryside', but they were carefully planned and managed.

The parks and gardens associated with great country houses and stately homes were specifically designed to convey a manicured image of the countryside and linked to the Picturesque movement in landscape painting. According to Mount (2013, p. 148), they were especially influenced by the paintings of Claude, whose 'Arcadian vision was copied across England in the parks of country houses, from Blenheim Palace to Stowe. Claude's pictures influenced Capability Brown's managed landscapes, which involved a natural look – albeit a contrived natural look. As Mount (ibid., p. 149) states, a 'Brown-style park attached to a country house became an essential feature, with thousands of them landscaped through the eighteenth and nineteenth centuries'; Brown's 280 commissions alone ended up covering more than 123,000 acres of Britain. Mount (ibid., p. 150) describes the various techniques applied to produce a contrived 'natural' look that was smooth and rounded, 'with vistas of intertwining hills and ornamental lakes'. These included clumps of exotic trees, originally planted in deer parks and then shifted, for scenic effect, to hilltops; hills often sculpted out of gentle slopes; lakes formed by damming streams; and serpentine lakes designed to cater for the taste for curved lines. Furthermore, by the 1750s Brown:

> took to bringing the park, thinly disguised as countryside, right up to the walls of the house by employing the use of the 'ha-ha' – a hidden ditch that kept animals from the house, but allowed views to run apparently uninterrupted, all the way from your front door to the horizon.

In establishing these real-fake rural landscapes, not only was the original land transformed, but also, across England, whole villages and their communities were removed and destroyed. At Stowe, for example, 32 houses and 180 people were removed to make way for William Kent's 'natural garden'. Also, where farming was involved as part of estate activities, great lengths were taken to keep the agricultural labourers and the less attractive aspects of farming out of sight. In fact, the excesses to which such landscape designers went to create their 'natural landscapes' evoked the comment from Tom Stoppard, in his play *Arcadia*, that a Brown landscape was 'manicured down to the right amount of sheep tastefully arranged' (ibid., p. 152).

While such practices no longer occur, many of the resulting landscapes, nevertheless, remain; some are still in private ownership, but many of the great estates are now owned and managed by the NT and their buildings and landscapes are maintained, and even reinstated, in their original appearance. Where estates involve active farming, tenant farmers are often required to farm using traditional methods; in Bransdale, North Yorkshire, for example, the NT paid for a modern barn to be enclosed within the façade of a traditional one so as not to spoil the image. The NT cares for 247,000 hectares of countryside, much of 'outstanding natural beauty', with nearly 20,000 hectares being parks and gardens; over 21 million visits were made to NT properties in 2014 (National Trust, 2014). Thus, not only do substantial numbers of people obtain a countryside experience involving these forms of landscape and scenery, but, in addition, such idyllic images are further perpetuated by the NT's merchandising operation with its prints, postcards, Christmas cards and calendars.

While country parks are, in many respects, very different places to the parks and gardens described above, they, nevertheless, share similarities. Like the landscaped parks and gardens, they are heavily managed and provide a countryside that is tamed, sanitised and safe. They are also easy to use in that great lengths have been made to provide for visitors' needs such as car parks, footpaths, seats, picnic tables, litter bins and a variety of other leisure facilities designed to enhance the visitor experience. While some country parks, such as Cannock Chase and Clumber Park, are very large, exceeding 1,000 hectares and containing some areas that might be described as semi-natural, the great majority of the 250 country parks in England and Wales are small or medium, with obvious management features.

Although country parks are clearly examples of cultivated countryside, for the millions of people who visit them they provide a perception of a real countryside experience (which, of course, is what they were designed, in part, to do). A vivid illustration of such perceptions was provided in the results of a survey of visitors to Priory Country Park, near Bedford, in the late 1980s (Bull and Bonser, 1988). In answer to a question that required respondents to indicate, using a five-point Likert scale, the extent to which they agreed with the statement: 'Priory Country Park is a place where you can enjoy true countryside', over 79 per cent said they agreed, with 44 per cent saying they strongly agreed. And yet this country park is located at the very edge of the town, with houses and other buildings being very visible – including, at that time, the cooling towers and chimney of the nearby coal-fired power station. Furthermore, the park contained all the ubiquitous features of country parks (footpaths, picnic tables, litter bins…), as well as a visitor centre.

An unwelcome countryside

In addition to the less welcoming aspects of village communities, mentioned in an earlier section, there are also restrictions on public access. Part of the perceived attraction of the countryside is that visitors can roam freely and, in some countries, this is broadly possible. In Sweden, for example, the practice of *allemansrätten* gives the public the right, under certain conditions, to stay temporarily on private land (Shoard, 1987). However, in the UK this is not the case and, for over a century, there has been a continuing debate about countryside access that has generated a sizable literature (e.g. Countryside Commission, 1988; Shoard, 1980, 1987, 1999; Curry, 1994; Watkins, 1996). Much of the issue concerns the private ownership of rural land and the reluctance of landowners and farmers to allow public access. In part, this reluctance was due to fears, real or imagined, that visitors would cause damage and should be kept away, but it has also been reinforced by successive government policy that, until recently, has not wished to challenge the power of landowners – one that was also linked to an elitist attitude among rural planners that visitors to the countryside should be strictly controlled (Curry, 1994). While various initiatives, and even legislation such as the 1990 Rights of Way Act, have attempted to improve access opportunities, such efforts have often been thwarted by farmers and landowners; it was not until the most recent legislation, the Countryside and Rights of Way Act (CROW) 2000, that significant progress was made, affording the public a legal right of access to approximately 4,000 square miles (*c.* 1 million hectares) of land, primarily in the uplands. Access remains a problem in much of lowland Britain, however, where there is evidence that many footpaths remain in a poor condition. The results of the Ramblers' Association's 'biggest-ever' footpath survey in 2015 revealed that '35% of paths [were] in need of improvement and nearly a tenth (9%) [were] difficult or impossible to use' (Ramblers' Association, 2016).

Implications of the rural idyll

While the idyllic image is clearly mythical, it occupies a prominent position in the minds of many and produces a powerful set of influences. It has already been shown how such a perception persuades substantial numbers of people to live in the countryside and how it is viewed by millions of urban visitors as a place 'of retreat and refreshment' (Aitchison *et al.*, 2000). But perhaps its most fundamental influence is its role in creating a powerful preservationist lobby and its effects on rural planning.

Much like historic cities, the countryside has been a preserved environment. Since the late nineteenth century a number of environmental/preservation organisations have emerged to campaign to protect the countryside from urban encroachment and development (Lowe and Goyder, 1983; Lowe, 1989) This broad political movement, supported in particular by the urban middle class, or at least that portion of the middle class that constitutes the 'service class' (Thrift, 1989; Urry, 1995c, 2002), involves such organisations as the NT, RSPB, Campaign to Protect Rural England, Friends of the Earth, Woodland Trust and County Conservation Trusts. In recent times the growth in membership of some of these groups has been massive. For example, the membership of the RSPB has grown from a mere 10,000 in 1960 to over 1 million members today (RSPB, 2016) and the NT's membership expanded to 4.2 million in 2014 (National Trust, 2014). These bodies provide substantial support for an urban containment policy that has existed since the late 1940s and has protected the countryside from considerable development (Hall, 1973; Gilg, 1996). Such a policy involved strict planning controls over development in rural areas – especially in designated areas such as national parks, areas of outstanding natural beauty, heritage coasts and green belts – helping sustain what Urry (2002, p. 98) refers to as the 'romantic gaze of the countryside'. Its downside, however, is its impact on limiting the supply of housing and affordable homes, the resulting rise in house prices, housing in urban areas being built in ever greater densities, its restrictions on industrial development and related employment and even its prevention of certain tourist developments such as hotels and holiday villages. Allied to these anti-development values are views about what constitutes appropriate forms of tourist activity in the countryside. The prevailing consensus amongst various rural policy makers is that only those activities that do not offend the countryside aesthetic are acceptable. Consequently, passive activities such as walking, sightseeing and watching wildlife, together with certain 'traditional' active pursuits are appropriate, whereas noisy sports or activities which attract large crowds are not.

Conclusion

The previous pages have examined popular perceptions of the countryside that are often described as idyllic. As discussed in Chapter 9, the romantic gaze overlays an image on actual locations and the imagination can be selective. While most authorities on rural matters may regard such views as fake, they, nevertheless, exert a powerful influence and this chapter has sought to explain why such perceptions are so entrenched within Britain's cultural consciousness, the reasons why they could be defined as fake-real, and the ways in which they exert such a powerful influence on the planning and management of the rural environment. The cultural warm authentication of the rural environment may seem anodyne and saccharine to some, but the chapter outlined how the construction became a part of the national psyche and collective memory through traumatic events such as industrialisation and the First World War. This may indicate why fictionalised versions of the countryside still exercise a visceral force on perceptions, leading to a search for authenticity and diffuse sense of loss.

It would, of course, be wrong to suggest that the countryside is entirely bereft of attractive landscapes or that the idyll is entirely a figment of the imagination. The millions of people who have moved to live in rural settlements or who visit the countryside as part of their leisure do not suddenly find they have made a major error and move back to the town or never visit the countryside again; in fact, surveys of rural tourism have shown that many visitors are making repeat visits. Furthermore, the millions of people who give their support to the various organisations that form the rural preservationist lobby clearly do so because there are substantial swathes of attractive countryside that they believe need to be protected. As Bill Bryson (2015, p. 33), a former president of the Campaign to Protect Rural England, remarks in his book, *The Road to Little Dribbling*: 'nothing – and I mean nothing – is more extraordinary in Britain than the beauty of the countryside'. But, as mentioned in earlier sections of this book, the villages in which people live are not the close-knit, friendly communities of the imagined stereotype, and the places that people visit have been manicured to produce a countryside more in keeping with the idyll than with the reality of a working, rural landscape. Much of the countryside is, in fact, a preserved landscape, protected by substantial financial support to farmers and by strict planning controls over development, especially in designated areas such as national parks and areas of outstanding natural beauty. For many people, whether the countryside they imagine and/or experience is real or fake is probably immaterial. However, what is real are the impacts such culturally constructed views have on its planning and management, and the fact that what are mainly the views of one class of society have an overwhelming influence on the lives of all those who live and work in rural areas, as well as those who live in urban areas in greater densities than they might otherwise need to.

At the end of Flora Robson's (1943) *Lark Rise to Candleford* the author describes walking through the invisible threads of spiders' webs, which she felt were trying to bind her to the village which she is leaving. In the same way, the rural idyll binds some of us with threads 'more enduring than gossamer' to search for a collective past in a place which may no longer exist, if it ever did.

8 Wilderness regions

From national parks to existential authenticity

This chapter acts as a contrast to the variety of man-made places discussed in the book, and focuses on wilderness areas, which are perceived as rare, singular and 'original' in a more-than-human world. Their aura of uniqueness and remoteness makes them objects in the search for authenticity (MacCannell, 1973), which has been discussed in previous chapters. Many of these places are regarded as pristine, natural environments, examples of 'objective authenticity' (Wang, 1999); however, a deeper examination reveals that 'wilderness' is a complex and contestable concept which may also be culturally constructed. The chapter unpicks the notion that they are, in fact, untouched, and cites ways in which they are managed and officially authenticated through designations, such as national park status.

The areas considered in this chapter possess unique qualities and are markedly different from the environments that people normally inhabit; their rarity and/or fragility often requires some form of protection. Such places are largely devoid of people, other than occasions when limited numbers of tourists are present, and their principal assets are their natural landscapes and/or wildlife resources, which are perceived to be largely unaffected by human agency and seem, in the case of the sublime, to have an agency of their own. In *Mountains of the Mind*, Macfarlane (2003) discusses the hold mountains have over him psychologically as archetypal symbols of freedom, individuality and ruggedness, which aligns with Wang's (1999) concept of existential authenticity. This has been indicated in films such as *Wild* (2014) or *Into the Wild* (2007), where the wilderness is viewed as simultaneously peaceful and serene, but also challenging, healing, hostile and dangerous, a place to test the mettle. The chapter places the spiritual and psychological pull of such places in the context of such movements as American Transcendentalism. In addition to the areas that most people would regard as wilderness, the chapter also addresses other scenic landscapes, which are often regarded as 'natural', but which involve various forms of management designed to protect them from either human impact or ensure certain valued wildlife is not adversely affected by natural ecological processes. It is the purpose of this chapter to consider these different places and to examine the idea of both naturalness and wilderness, and the extent to which they are contested concepts.

The concept and nature of wilderness

Wilderness areas are generally accepted as those places that are remotely located beyond the inhabited and managed countryside and which possess what many would regard as 'natural ecosystems'. They are places 'on the edge', usually perceived as being wholly, or at least largely, uninfluenced by human activity and include such environments as mountains and remote uplands, tropical rainforests, polar regions, areas of tundra and taiga, and deserts. Despite the concept of wilderness encompassing the idea of naturalness, according to Dudley (2011) the idea that such places still exist is a myth, as there is nowhere left on earth that is wholly uninfluenced by human activity of some kind. For example, pesticides have been shown to affect species not simply near where they are used but much further afield because of various stages of ingestion through the food chain. Also, various airborne pollutants can have widespread impacts, such as acid rain resulting from industrial activity in Western Europe destroying trees in the relatively remote northern coniferous forests. Furthermore, the effects of greenhouse gases on both the ozone layer and average world temperatures are on a global scale and, thus, even the most remote areas of the polar regions and high mountain ranges are not immune from impact. However, it is not just these more recent and, perhaps, more well-known influences that have impacted supposedly wilderness areas. Various impacts and pressures have been altering ecosystems, either intentionally or accidently, for thousands of years. Dudley (ibid.) cites several examples of areas that are perceived as natural, but which have been affected by incursions of new settlement – especially colonial expansion in the eighteenth and nineteenth centuries. Even earlier examples can be found in relation to some of the world's best-known 'natural forests' around the sites of former civilisations such as in Central America, South East Asia and the Amazon. 'Throughout much of Central and South America, for instance, "natural forest" grows over the ruins of earlier Mayan, Aztec, Toltec and Inca civilisations that at one time had cleared large swathes of land' (ibid., p. 11). Nevertheless, although such areas can revert to essentially natural ecosystems, it has been shown that such forests possess markedly different and less diverse tree composition (ibid.). In addition, there are various examples of indigenous populations changing the nature of ecosystems through management regimes such as managing forest and grassland areas through systematic burning, culling wildlife, especially large mammals, and encouraging fruit-bearing plants. And today, of course, many well-known wilderness areas are officially authenticated through their national park designation and thus subject to management regimes that maintain, or in some cases transform, their characteristics in ways that society considers appropriate. It is clear from the above discussion that the concept of 'wilderness' is not clear cut and, as hinted at in the last sentence, it is further complicated by cultural components. As Cronon (2003, p. 214) has stated 'it is quite profoundly a human creation' and he develops his argument thus:

> It is not a pristine sanctuary where the last remnant of an untouched, endangered, but still transcendent nature can for at least a little while longer be

> encountered without the contaminating taint of civilization. Instead, it is a product of that civilization, and could hardly be contaminated by the very stuff of which it is made. Wilderness hides its unnaturalness behind a mask that is all the more beguiling because it seems so natural.

Although modern dictionaries define wilderness as uncultivated and otherwise undeveloped land with an absence of humans and presence of wild animals assumed, Nash (1967, p. 1), in his text *Wilderness and the American Mind*, argues that a universally accepted definition of 'wilderness' is elusive due to its subjective nature, being 'so heavily freighted with meaning of a personal, symbolic, and changing kind'. In examining the etymology of the word as developed in the early Teutonic and Norse languages, from which the English word, in large part, originates, he explores the various roots of the word and suggests it derives from the Old English word *wildeor*, meaning 'of wild beasts' ('wild' used to convey the idea of being lost, unruly, disordered, or confused, and Old English *deor*, meaning animal). Wilderness, then, becomes 'the place of wild beasts'. However, one of the problems of defining wilderness refers to the ways in which the word is expressed in various European languages, with the Teutonic view being derived from a fear of the vast northern forests, whereas in the Romance languages of southern Europe it is influenced by concepts of solitude, emptiness, loneliness and lack of civilisation (Nash, 1967; Hall and Page, 2006). Tuan (1974) describes how both the north European and Mediterranean traditions define and portray wilderness as a landscape of fear, which is outside the bounds of human settlement. This idea is also noted by Nash (1967, p. 1) who cites the example of the eighth-century, epic work, *Beowulf*, where the word *wildeor* 'appeared in a reference to savage and fantastic beasts inhabiting a dismal region of forests, crags, and cliffs'. However, Tuan (1974, pp. 109–10) also suggests that in the Bible the term recalls two contradictory images. On the one hand, it is viewed as 'a place of desolation, the unsown land frequented by demons' and condemned by God – as Nash (1967, pp. 14–15) suggests, a place symbolised as 'an environment of evil, a kind of hell' in which the arid wasteland was identified with God's curse. On the other hand, it can be seen as 'a place of refuge and contemplation' and 'any place where the Chosen are scattered for a season of discipline or purgation' (Tuan, 1974, p. 110). Tuan also illustrates how the ascetic tradition in Christianity maintained the dual and opposed meaning of wilderness and how its ambiguity also extended to America, where the New England Puritans believed that their reformed church would blossom like a garden in the protective wilderness, but where some believed wilderness was the place 'where nothing appeared but hard labour, wants and wilderness-temptation'.

Macfarlane (2008, p. 30) also refers to the 'two great and conflicting stories' that have been told about wilderness: that it is a quality to be vanquished, but also a quality to be cherished. The idea of taming and cultivating wilderness, of little value in its original state, has been a ubiquitous view in cultures, ancient and modern, Eastern and Western; but there is also a parallel view of wilderness having a particular energy with wild places being seen as 'realms of miracle, diversity and abundance' (ibid., p. 31). Macfarlane also refers to the writings of monks in late Saxon times, which praised the beauty and 'riotous fecundity' of the wild, and cites similar evidence of a love for the wild in the Chinese artistic tradition of *shan-shui*.

The different ways in which religious thought has influenced the perception of wilderness is part of the cultural construction of the concept and this construction has changed through time. In his seminal work *Topophilia*, Tuan (1974) provides a detailed examination of people's changing environmental attitudes towards mountains that includes examples from different cultures throughout the world. He shows that in both Eastern and Western civilisations changes involved shifts 'from a religious attitude, in which awe was combined with aversion, to an aesthetic attitude that shifted from a sense of the sublime to a feeling for the picturesque; to the modern evaluation of mountain as a recreational resource' (ibid., p. 71). Evidence of an aesthetic appreciation of mountains can be traced back to the Sung dynasty (AD 960–1279), whereas in the West it came much later. Here aesthetic aspects of wilderness have become established over the last couple of centuries and, according to Tuan (1974), the change involved various factors that included a rejection of the long-held view that circles and uniform features represented perfection, the popularity of Chinoiserie and Chinese landscape design, easier forms of travel that made mountainous regions more accessible and ideas about the health-giving properties of mountain air.

In the West the view of mountains as unsympathetic places with few people finding beauty in wilderness settings prevailed until the nineteenth century. As Boswell commented in his *Journal of a Tour to the Hebrides*, published in 1785, 'A landscape or view of any kind is defective… without some human figures to give it animation' (quoted in Withey, 1997, p. 38). Evidence of cultivation and habitation made a landscape more welcoming to the eighteenth-century eye; vast open spaces, dense forests or rocky terrain were likely to be criticised as monotonous and unpleasant (Withey, 1997). Prior to this time, Europeans thought of man as the centre of the world and nature as something to be shaped or exploited for his purposes (ibid.). Gradually, however, this attitude changed towards one that valued the natural world and found beauty in it; such changes in attitude were also linked to the cultural changes stemming from the Romantic movement, in part linked to religious beliefs. As Withey (ibid., p. 39) points out:

> Christian doctrine, going back to Augustine, held that God had produced the irregular and barren features of the earth after the original creation, in his wrath at Adam's fall; thus it was possible to believe in the perfection of God's creation while detesting some of the earth's physical forms, which were meant to be ugly and horrifying. In the late seventeenth and early eighteenth centuries, however, a new generation of theologians began to argue that the earth in all its forms was part of God's original design and that the greatness of God was manifest in nature.

The North American experience provides a classic example of these changes in attitudes, as illustrated by white Americans' relationship with 'the West' (Shaw and Williams, 2002). Early white settlers believed the great grassland plains of the continental interior to be 'deserts' and of no agricultural value as they were devoid of trees and thus alien to the European experience (Watson, 1970). They quickly crossed such areas as they headed for California and Oregon, but beyond the plains they then encountered the rugged and mountainous West as an area beyond human control (Hartmann and Hennig, 1989, cited in Shaw and Williams, 2002). Such a view contrasted markedly with that of the native Americans who had inhabited the region for millennia and whose relationship with nature was more in line with that found in Eastern religions. However, in the mid- and late nineteenth century attitudes to the American West began to change; European settlers began to view the 'wilderness' of the frontier region in a more positive light, highlighting its aesthetic qualities. Influenced by ideas stemming from the European Romantic movement, it also came to be viewed as an important contrast to urban-industrial areas and capable of providing restorative and psychological benefits to visitors. Ralph Waldo Emerson and Henry David Thoreau encouraged the doctrine of self-reliance in books such as *On Walden Pond*, associated with finding the authentic self in nature. The concept of the Sublime was promoted by the Transcendentalist painters from the Hudson River School in the mid-nineteenth century such as Frederic Edwin Church, who depicted the monumental scenery of America. This was the rural idyll as seen through a wide-angled lens, a spectacular, sublime hot authentication of place designed to draw people westward.

This cultural influence also became associated with America's political development, with wilderness symbolising the 'frontier' and pioneering spirit (Gold, 1980). Cronon (2003, p. 200) links such ideas to 'the powerful, romantic attraction of primitivism – the belief that the best antidote to the ills of an overly refined and civilized modern world was a return to a simpler, more primitive living' and 'in the United States, this was embodied most strikingly in the national myth of the frontier'. It was not just the symbolism of frontier, however, that linked wilderness to American political development and the concept of Manifest Destiny, which encouraged pioneers to 'Go West'. As Hall and Page (2006, p. 257) point out 'the political independence of the American nation found cultural expression in the extolment of the virtues of American natural scenery', especially in the fact that it was a pristine environment 'untainted by the domination of things European'. Runte (1997, p. 11) provides an explanation of this development in that the USA had suffered the embarrassment of a dearth of recognised cultural achievements, for – unlike European countries which traced their origins far back into antiquity – it lacked a long artistic and literary heritage and reminders of a human past in the form of castles, ancient ruins and cathedrals. However, by the 1860s many thoughtful Americans had embraced the West as a replacement for man-made marks of achievement. The agelessness of monumental scenery, such as the Rockies and features like the Grand Canyon, instead of the past accomplishments of Western civilisation, was to become the visible symbol of continuity and stability in the new nation.

The changing attitudes to wilderness also became the foundation of the movement to create national parks in the USA, although the rationale for the establishment of Yellowstone, the first national park, in 1872, did not take account of the aesthetic, spiritual, or cultural values of wilderness, but, instead, was concerned with preventing private acquisition and exploitation of geysers, hot springs, waterfalls and similar curiosities (Nash, 1967, p. 108). In fact, initial public reaction to the creation of Yellowstone National Park also ignored wilderness; instead 'it was praised as a "museum" and "marvellous valley", an area where people could see the "freaks and phenomena of Nature" along with "wonderful natural curiosities"' (ibid., p. 113). Gradually, however, Congress realised that Yellowstone National Park was a wilderness preserve; further developments were aided by the campaigning of key individuals like John Muir, the establishment of preservationist organisations such as the Sierra Club (established in 1892) and, later in the twentieth century, the recognition of the scientific and economic value of protecting wilderness, as well as its importance for tourism (see section below). A further development later in the twentieth century was the enactment of important legislation, namely the Wilderness Act of 1964, designed to protect many areas where the imprint of human agency is largely unnoticeable. Today, there are about 40 national parks in the USA, mostly in the West and Far West, with other such designations throughout the world in countries such as Canada, New Zealand and Australia (see Hall and Page, 2006, pp. 258–9, for a comprehensive account of key developments in the establishment of national parks in these four countries).

Wilderness and tourism

Central to the view of the wilderness as an authentic place, which brings out the authentic self, is the recognition of the aesthetic qualities and potential psychological benefits. This is part of a further cultural construction of wilderness – namely its importance for tourism – and, once again, the North American experience provides a vivid illustration. The decades following the Civil War saw increasing numbers of wealthy Americans seeking out wilderness as a tourism destination and recreational resource. Thousands of urban dwellers from the East came to gaze in awe at the impressive 'monuments' of the Rockies (see Figure 8.1) and its unique wildlife, with Yellowstone National Park attracting over 5,000 visitors a year after the railroad arrived in 1883; it has been said that Yellowstone was established as a destination because of the railroad. Apart from establishing the idea that wilderness was now an attraction, even a source of entertainment, tourism also introduced additional aspects to the concept. The idea of wilderness as a non-peopled place was a factor that led to the removal of indigenous Indian tribes from some of the most attractive wilderness areas. Native Americans had lived in harmony with nature, were the ultimate predators within the natural ecosystem and even managed certain wildlife through controlled burning and encouraging food-bearing plants. They were, thus, an essential part of the authentic environment, but this was either ignored or not understood; in any case, it did not accord

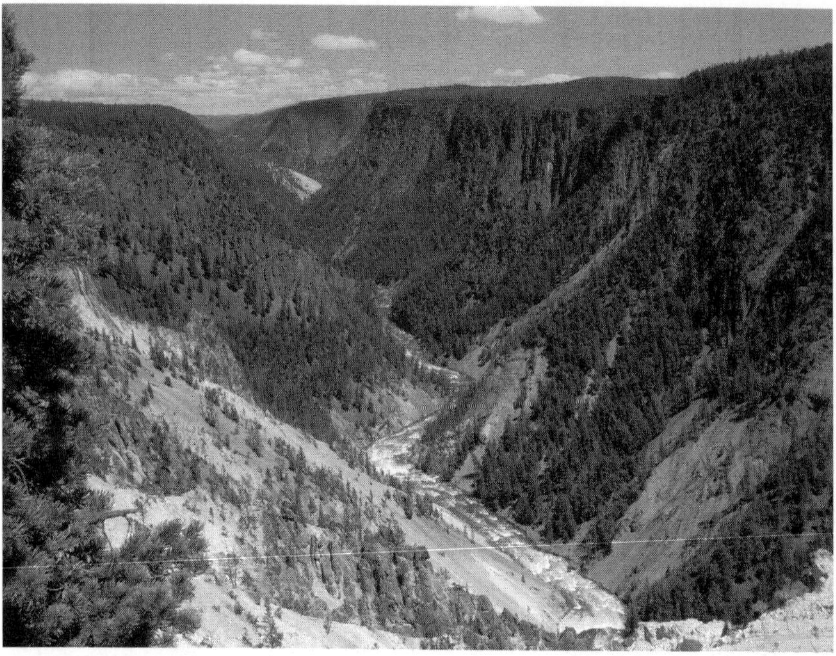

Figure 8.1 Grand Canyon of the Yellowstone: Yellowstone National Park
Source: Author

with the tourists' perceptions. As Cronon (2003, p. 224) points out, 'the removal of Indians to create an "uninhabited wilderness" – uninhabited as never before in the human history of the place – reminds us just how invented, just how constructed, the American wilderness really is'. As others have also pointed out, it is 'an artificial construction that meets people's needs… a kind of domestic wilderness, but it's hardly the authentic real place that it purports itself to be' (BBC, 2011).

Allied to this, although in some ways a contrary notion, was the idea that the wilderness that people wanted to visit needed to be a tamed, gentler sort of wilderness, rather than one that was 'red in tooth and claw', and this required the application of specific management interventions. One consequence of this in the Yellowstone National Park, for example, was the culling of large predators such as bears and wolves such that, by 1926, wolves had been completely eradicated. The science of ecology was still in its infancy and few understood the natural processes operating within such an ecosystem. With their natural predator no longer present, elk numbers increased substantially and, although, like bison, they were popular with tourists, their increasing numbers resulted in the over-grazing of vegetation such as willow and aspen. Eventually, to prevent further destruction of such tree species, the elk population had to be culled and, finally, in 1995, 70 years after their departure, wolves were reintroduced to establish the 'natural' control mechanism for the continuation of this 'wilderness' ecosystem (ibid.). The popularity of Yellowstone for tourists and their interest in particular forms of wildlife in the later nineteenth and early twentieth century also led to other aspects of inauthenticity. The National Park Service, established in 1916, did what it could to ensure that tourists got to see animals at close quarters, so animals were fenced in and fed and there was also a captive breeding programme for bison. Visitors driving through Yellowstone, and observing bison in the distance, would assume they were seeing wild, native animals but discreetly concealed was the fence that kept them in view (ibid.).

Cultivating an authentic notion of wilderness for tourism has involved several different aspects. The attraction was initially entirely a passive one, of sightseeing, as it still is for many people today. Apart from those tourists who simply came to look and marvel at the uniqueness or grandeur of the landscape and its wildlife, wilderness also attracted those who regarded wilderness areas as challenging environments in which to pursue a variety of active forms of recreation and sport, and those who sought an opportunity for escapism and spiritual experiences, for losing oneself in an environment devoid of people. Such life-enhancing experiences can be seen to equate with Wang's (1999) concept of 'existential authenticity', thus adding further complexity to the authentic nature of wilderness.

The importance of wilderness areas for sightseeing relates, in part, to the earlier discussion about the changed perception of it and the way in which many people came to value wilderness as an important and legitimate focus of the 'the tourist gaze' (Urry, 2002). However, the changing aesthetic value of wilderness was not only linked to changing religious attitudes, but also to a reaction against nineteenth-century urban industrialism. Wilderness tourism thus became a means of escaping from the polluted air and crowded, dirty and insanitary towns and cities that were growing apace in the nineteenth century. Such sentiments led not only to increasing numbers of people wanting to holiday in such areas – in the USA the railroads and, in the twentieth century, the motor car made wilderness areas more accessible – but also to systems of planning and environmental protection and the strengthening of the case for national park designation.

Other wilderness areas have also been transformed in different ways – like the European Alps, which have been reconstructed into sports tourism places (Withey, 1997; Beedie and Hudson, 2003; Hudson, 2003; Weed and Bull, 2009). The attraction of the Alps first developed as a later part of the Grand Tour, when the earlier 'classical Grand Tour', linked to galleries, museums and high-cultural artefacts, eventually shifted to the 'romantic Grand Tour', which saw the emergence of 'scenic tourism' and a new taste for mountain scenery (Towner, 1996). This change in tastes, the development of centres in Alpine regions and transport developments that made such places ever more accessible thus paved the way for the subsequent growth of sports activities such as climbing and skiing which are so prominent today. Visiting glaciers has also gained a nostalgic feel as tourists measure their visible retreat in places such as the Perito Moreno glacier in Patagonia. Mountaineering, while not discussed in this chapter, has also created unsustainable aspects such as the Coca-Cola Trail on Mount Kilimanjaro, where litter and crowding from the sheer number of mountaineers was affecting the environment and safety. The sheer popularity of mountaineering, as demonstrated in the film *Everest* (2015), as people seek the aura of the untouched, pristine place and ultimate 'bird's eye view', has threatened the existence of some areas.

Also, in more recent times, perceptions of wilderness have been reinforced by marketers who advertise wilderness and national park areas as tourist attractions. Mowforth and Munt (1998, p. 166), for example, refer to 'wilderness' as 'a powerful notion that is repeatedly conjured up in tourist media such as travel brochures' and such views are further reinforced by the number of television programmes devoted to aspects of wilderness – some purely educational, but others concerned with survival in such extreme environments. In addition, the growth of ecotourism/nature tourism and 'alternative' holidays is also influential in emphasising the tourist importance of wilderness areas. Such forms of tourism have become ever more popular among the middle classes, those with lifestyles strongly motivated by creativity, health, new experiences and personal growth (Gordon, 1991, cited in Shaw and Williams, 2004), who regard them as the ultimate contrast to the mass tourism associated with cities and seaside resorts. For these groups, despite the reality, wilderness areas possibly do fulfil their 'search for the authentic'.

While wilderness areas have become commodified for tourist enjoyment, the more tourists are attracted to them, the less authentic they become. The fact that tourists inevitably produce impacts on such pristine environments is one of the essential truisms of tourism activity – not only do tourists produce visual and ecological impacts, their presence may also lead to staged authenticity. A good example of this involves Hurtigruten cruise holidays to northern Norway in the winter months and their marketing literature refers to the Arctic wilderness where, in addition to viewing spectacular scenery and natural wonders such as the *aurora borealis* (Northern Lights), travellers can experience dog sledding and a reindeer sleigh ride through the snow-covered landscape. One such reindeer excursion involves visiting a small Sami encampment where the reindeer are kept (see Figure 8.2); in addition to the ride itself, visitors are given an opportunity to show their skill at throwing a lasso at reindeer antlers and are welcomed into a *lovvu* (the temporary tent-like dwelling that was used by the Sami to follow the reindeer herds), where they can sit round an open fire and be served the traditional Sami meal of *Bidos* (reindeer broth).

While giving the tourist a sense of what life for the Sami might have been, the experience is a staged one. It is unlikely that these Sami reside permanently in such dwellings and structures, but, more likely, live in the nearby city of Tromso; and, rather than pursuing the traditional way of life of their forebears, instead they are rather part of a modern tourism industry.

The British experience of wilderness

In Europe, unlike the other continents, there were virtually no wilderness areas left by the nineteenth century. Nevertheless, the more remote, marginal areas of

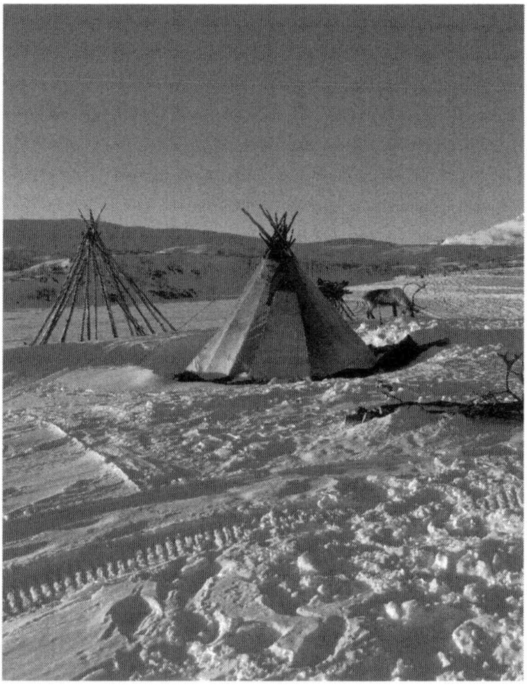

Figure 8.2 Sami camp
Source: Author

the continent have also undergone a process of cultural reconstruction like that experienced elsewhere. In Britain, for example, various upland environments that were once perceived to be dark, dangerous and uncivilised, and to be avoided if possible, have, in the last couple of centuries, acquired a very different perspective. While not truly wilderness – or even near wilderness, in contrast to the areas that have been discussed in the previous section – they are often perceived as such or at least viewed as semi-wilderness and, like wilderness areas elsewhere in the world, have attracted the protection of national park status. In the UK, for example, the 1949 National Parks and Access to the Countryside Act led to the subsequent establishment of ten national parks in England and Wales that were predominantly in upland regions. They were established to protect their scenic and wildlife qualities and restrict urban development, but also to provide opportunities for recreation and tourism. However, unlike those in true wilderness areas, the landscapes involved are essentially working landscapes and have clearly been influenced by human activity; most of the land is privately owned rather than owned by the state. As McEwen and McEwen (1982) have remarked, in many respects they are neither national nor parks.

Another key difference between most British wilderness areas and those in North America, where the principal areas include primeval forest, is the importance of moorland which Rackham (1986, p. 305) detailed as by far the 'most extensive natural vegetation running throughout Highland regions (though not necessarily at high altitudes) from Land's End to Cape Wrath and the Shetlands'. While initially referring to it as 'natural', Rackham (ibid., p. 306) also examines the debate as to how far moorland is in fact 'natural'. He cites the tradition of literary writers who 'suppose that moorland is primeval wilderness', but reports that 'in recent decades the opposite theory has gained ground, that all moorland is an artefact made by the destruction of wildwood'. He cites the evidence of trunks and stumps of trees that have been dug out of bog peat in various moorland environments and a study based on pollen analysis by Turner and Hodgson (1979, cited in Rackham, 1986, p. 306), which showed that 'in Mesolithic times wildwood covered the present moors of northern England right up to the highest summits of the Pennines'. However, he also reports the findings of another study by Birks and Madsen (1979) where 'pollen analysis shows that much of the moors of Shetland, Orkney, the Caithness plateau and the Outer Hebrides is natural tundra which has never had trees'.

In terms of their naturalness, Shoard (1982) also points out that upland environments are not the oldest landscapes in Britain, suggesting that heath communities were created approximately 4,000 years ago because of the destruction of forest to provide wood, charcoal and sheep runs and have been maintained by subsequent human activity involving periodic burning and sheep grazing. More recently, this has been important as part of the management of the moors for grouse shooting. Burning kills off the older, woody heather and allows new plants to grow, providing young shoots for the grouse to eat. Were the heather allowed to grow naturally it would not produce the short, springy plants that characterise the modern landscape and would eventually be replaced by more dominant birch scrub or woodland. While the purple heather moorlands are generally regarded as a particularly attractive feature of many upland areas, such ecosystems are not rich habitats; in fact, some ecologists have gone so far as to label moors as 'biological deserts', compared with habitats such as deciduous woodland and chalk grassland (ibid., p. 4). Such upland environments do not have the same ecological or 'natural' significance that wilderness areas elsewhere in the world possess; yet upland heath and the other upland community, blanket bog, still contain some unique and relatively rare species such as sphagnum moss, alpine plants that are relics of the ice age and certain bird species that are either adapted to the specific moorland ecosystems such as grouse and curlew or others, especially certain raptors such as the hen harrier, that have been confined to the margins as a result of human persecution.

While not rich in terms of diversity, such communities are therefore valuable in terms of their wildlife and this value is highlighted by Mabey (2000), who describes the unique environment of the Flow Country, 1,500 square miles of wild peatland in Caithness and Sutherland. The area is the world's largest remaining concentration of blanket bog, a series of rain-drenched plateaux covered by sphagnum moss and pools that contains one of the richest collections of breeding birds in Europe such as dunlin, greenshank and golden plover. Mabey (ibid., p. 163) originally wrote his account as part of a journalistic assignment during the late 1980s when the Flow Country, which he describes as 'a kind of tundra', was threatened by plans to turn a large part of it into commercial forestry. The forestry interest regarded the area as a wasteland with little in the way of productive capacity and thus, so it was argued, forestry (through the help of generous tax breaks) would be helping to 'improve' the area and produce some form of economic return. The argument clearly saw wilderness in pejorative terms, as an area needing to be tamed, very much in line with the negative tradition discussed earlier in the chapter. The conservationist interest, opposing such developments, did not see such a place as a wasteland, but instead viewed the wilderness in terms of its valuable wildlife and as a special place to be protected (it now has special protected status). Once again, it is clear that the nature of wilderness involves contradictory perceptions.

Other writers acknowledge the contrasting attitudes towards moorlands. Atkins (2014) has explored in considerable detail the nature and characteristics of the moorlands of England in his book *The Moor*, sub-titled *A Journey into English Wilderness*. In it he describes moors in various parts of the north and south west of the country and cites various eighteenth-century writers and philosophers who regarded moorlands as inferior to both the 'pastoral valley' and the mountains in terms of attractive landscape. He sums up this discussion thus:

> While God resided in the mountains, and social man in the valley, the moor was where the outcast went – the fugitive, the savage, the misanthrope. 'The wild moors' could have 'no interest' for strangers, wrote Charlotte Bronte. Moorland, it seemed, was merely a 'desert' that you had to slog across to reach your destination. In the valley you sighed; on the mountain top you gasped. On the moor you merely held your breath.
>
> Ibid., p. xix

But, despite also feeling hesitancy about heading out across various stretches of moorland – 'the old, animal revulsion induced by hostile terrain' – Atkins (ibid., p. xx) still experienced an alluring attraction to them. For him, the moor 'promised something that the meadow, the pasture and the woods [could] not – if not 'heroic adventures' (the words of W.H. Auden), then a kind of reply to the 'portion of [himself] that remained uncultivated'. Clearly, for Atkins, the moor provided opportunities for existential experiences and thus he would 'take himself off' and explore the ten different areas of moorland that constitute the basis of his text.

Mark Tully (2010, p. 329), in a contribution to the *Icons of England* collection entitled 'Alone on the Moors', records similar sentiments. He talks of moorlands conveying 'a sense of the grandeur of nature, so much greater than anything… humans have created', but he also describes their bleakness: 'a humbling bleakness, which often gives a feeling of being small and insignificant'. Macfarlane

(2008) also develops the idea of bleakness and how people struggle to assimilate flat terrains and use terms such as 'stark', 'empty' and 'limitless' to describe them. He also considers the concept of vastness. Of the various wild places he explores, and in this case traversed, it is Rannoch Moor in the west of Scotland that he uses as his example of a moorland environment. Before crossing the moor on foot he had driven across it at night and had been surprised by its extent, which seemed to go on for 'mile after empty mile… as though [he] had driven into a pool of black and limitless space, and were passing through another, not altogether earthly, place' (ibid., p. 66). He also remarks, of his trek across the moor, on the way the moor's vastness and self-similarity affected his perception of distance. He returns to this theme in a more recent book, where he describes the moor of Lewis resembling 'a *terra nullius*, a nothing place' which is vast, flat and repetitive in form (Macfarlane, 2015, p. 16); he likens it to MAMBA country, the acronym used by Welsh farmers to describe the hills of the Elan range in mid-Wales: 'Miles and Miles of Bugger All'. But, despite what many would no doubt regard as a foreboding place, Macfarlane, like Atkins, was attracted to the vast expanse of Rannoch Moor because it was a rare, contrasting environment to the one in which he spent most of his life. As he says, the moors of Britain and Ireland have become wild places for numberless people, who leave behind the confines of their cities to enter another realm (Macfarlane, 2008, p. 80). In his 2015 text (p. 59), he describes Nan Shepherd's poems about the Cairngorms, which she believed contained an 'uncanny atmosphere whereby the hills are both hostile and habitual, unsettling and enfolding' and tells of his own experiences of walking in the Cairngorms and enjoying 'an astonishing time of gifts' (ibid., p. 79).

As Atkins (2014, p. 77) suggests, the attractions of the moors are difficult to explain but unmistakeable to experience. However, Shoard (1982, p. 4) posed the question of what exactly it was that defined wilderness for a group of people in England and Wales who seemed to associate the idea with moor and mountain. She concluded that those attracted to such landscapes seemed to be:

> seeking a context for the pursuit of their individual identity away from the herd. To do this they [needed] to get away from the environment their fellow men 'had' created… to a place as devoid as possible of what is obviously human handiwork. It apparently does not matter if the landscape is in fact man-made… so long as it looks 'natural'.

Shoard (1982) also identified several primary conditions that needed to be met if an area was to be considered 'wilderness' in a British context, together with secondary characteristics which, although not essential, may heighten the appeal. Primary conditions included wildness, openness, asymmetry, homogeneity, height, freedom for the rambler to wander at will and the absence of what is obviously human handiwork. Secondary characteristics included the possession of relics of ancient man, undulation, wind and the absence of human beings not fully appreciative of the role of 'wilderness'.

People's perceptions of British upland environments have also been influenced by tourism marketing. A number of writers have drawn attention to the way in which the Scottish Highlands have been culturally constructed as both a picturesque tourism destination and symbol of nationalism (Butler, 1985; Aitchison *et al.*, 2000; Shaw and Williams, 2002). The description of the Highlands has been created and continually reinforced through various writings (novels, poems and travelogues), paintings and TV and film (for a detailed coverage see Aitchison *et al.*, 2000), but, in some people's eyes, they are an imaginary creation in which 'an untamed and hostile terrain, peopled by war-like clans, became, through physical suppression, clearance and mythical re-imaging, a playground' for tourists (ibid., p. 92). The Highland landscape remains one of Scotland's strongest tourism images and, although no national parks were designated in the Scottish Highlands in the 1950s because such areas were regarded as too remote and thus not in need of such protection, their sustained popularity and improved access has led, more recently, to national parks being established in Loch Lomond and the Trossachs in 2002 and the Cairngorms in 2003.

While moorlands are the most obvious, and certainly most extensive, area of perceived wilderness in Britain, some writers have referred to certain other places in such terms. For example, Macfarlane (2008) lists the vast shingle peninsulas that jut from the coastlines of East Anglia and Kent (e.g. Blakeney Point in Norfolk, Orford Ness in Suffolk and Dungeness in Kent) as wild places, as well as saltmarshes, estuaries and forest. On a global scale, forest is perhaps the ultimate example of wilderness, with its very etymology linked to the association of wood and wild, but, although there is no true wildwood left in Britain, there are some forested areas that may still provide a semblance of such places. For example, Macfarlane (ibid., p. 89) talks of his visit (walking and camping) to Black Wood (to the east of Rannoch Moor), where 'at times the forest was so thick that any sense of direction came only from the sense of slope' and 'often, the only noise [he] could hear was the creak of boughs rubbing against each other in the wind'.

It can be argued that authentic wilderness can no longer be found in Britain. As Macfarlane (ibid., p. 127) points out:

> no such chaste land exists… and no such myth of purity can hold. Thousands of years of human living and dying have destroyed the possibility of the pristine wild. Every islet and mountain-top, every secret valley or woodland, has been visited, dwelled in, worked, or marked at some point in the past five millennia.

Nevertheless, wilderness is a relative concept and there are clearly areas in Britain that many people would regard as possessing key elements of wilderness. An internet search for 'British wilderness' reveals a host of sites with suggestions of places that accord with the description, many in the northern highlands and islands of Scotland but others in the upland areas of England and Wales.

Another, but very different, type of place that has also been described as wilderness is what Shoard (2000, p. 74) refers to as 'edgelands' – areas where the town and countryside merge or, in her words, 'the interfacial interzone between urban and rural'. Unlike the coverage in most of this chapter, where wilderness has been assumed to involve areas devoid of human occupation or at least where it is minimal, edgelands contain substantial evidence of human activity. As Shoard (2002, p. 130) describes, it is an area 'characterised by rubbish tips and warehouses, superstores and derelict industrial plant, office parks and gypsy encampments, golf course, allotments and fragmented, frequently scruffy farmland'. And yet, despite this, Farley and Symmons Roberts (2011) refer to it as 'England's true wilderness'. The reason for such a claim lies in its strangeness, its mystery, the fact that it is neither town nor country, and its appearance as 'a vaguely menacing frontierland' (Shoard, 2002, p. 130).

A final aspect that needs to be raised in relation to Britain's wilderness environments is whether certain areas of the uplands, moorlands and forests might be allowed to revert more to their original character and for natural ecological processes to proceed unhindered. This idea has been referred to as 'rewilding' and described as a 'strategy for the conservation of complete self sustaining ecosystems, primarily involving the protection and, where necessary, reintroduction of populations of keystone species in large, connective reserve networks' (Brown *et al.*, 2011, p. 288). Brown *et al.* consider its possible application in Scotland and suggest that 'not only is rewilding possible in Scotland, but it may be expected to provide numerous ancillary benefits' (ibid., p. 391) such as the further development of nature-based tourism and an increase in employment linked to rewilding projects. Rewilding has also been promoted by George Monbiot (2014) and the charitable organisation Rewilding Britain, whose mission includes 'the mass restoration of ecosystems'. While Brown *et al.* (2011) see rewilding as part of a broader conservation strategy, Monbiot adopts a more extreme stance, arguing that humans should step back from nature and let it go 'feral', as the title of his book suggests. He considers that many ecosystems are kept in a 'state of extreme depletion' and now involve 'the merest scraping of what were once vibrant and dynamic ecosystems' (Monbiot, 2013). While some conservationists may have sympathy with Monbiot's ideas, there are a number of qualifications and counter arguments. First, if natural ecological processes are allowed to proceed unhindered, the end result will not be an entirely natural, authentic ecosystem because such areas have already been affected by human activity and the resulting biodiversity will not be quite as diverse as the original primeval state. Second, certain wildlife communities are managed in order to preserve particular forms of wildlife and ecosystems that people value. Allowing certain areas to return to their original state would involve most non-farmed land reverting to woodland. While this might lead to an increase in certain woodland species, other species which benefit from managed areas would decline or even disappear, as would other valued ecosystems such as reedmarsh, heather moorlands and grasslands. As a comment on Monbiot's ideas from the Wildlife Trusts (the parent body of the County Wildlife Conservation Trusts) (2017) points out:

> Around half of our terrestrial species live in open habitats which require some sort of natural disturbance (or management intervention by humans) to exist – and which ultimately would be at risk of extinction with no type of management at all.

Conclusion

This chapter provides a contrast to many of the discussions of authenticity and inauthenticity in more overtly man-made places examined in the book. The previous pages have examined the nature of wilderness and the related concept of naturalness and such ideas are both complex and dynamic examples of 'emergent authenticity'. The prevailing consensus and common dictionary definitions would suggest that wilderness is an undeveloped area that is free from human activity. However, various authorities have shown that few, if any, such natural ecosystems exist and that many of the classic wilderness areas of the world were inhabited by humans for thousands of years. This provides the ultimate example of the negotiability of authenticity. Such peoples lived in harmony with their environments, even modifying the natural processes of such ecosystems through various management interventions. Far from viewing such places in negative terms, as European settlers originally perceived wilderness, they were instead viewed as benign environments that provided their human inhabitants with all their essential requirements for life.

Wilderness, as it tends to be understood today, is essentially a cultural construction but also a construction that has changed through time as ambivalent religious attitudes have given way to various Romantic notions and to wilderness being important for leisure and tourism. Even some of the very characteristics of wilderness that are often used to describe it as a foreboding place – remote, desolate, hostile – have, for some, become the very attributes that make wilderness areas attractive. Wilderness areas provide a set of resources for increasing numbers of people who want to pursue a range of activities hotly, warmly and cooly authenticating their environment; many who simply want to see the unique landscapes and feel a sense of the sublime in a safe, managed fashion, and others who want to pit their skills against the wild environment that such places may offer. For both groups, however, such places provide unique experiences and, thus, such areas, although not entirely authentic, may well provide examples of Wang's (1999) 'existential authenticity'.

The importance of such places can be seen in the many areas throughout the world that have been given protected status; even in heavily populated regions like Britain, the value of places perceived as wilderness is also recognised through the 'official authentication' of national park designation. Such environments provide a semblance of wilderness for people who are not necessarily able to travel to the far-flung regions of the world in pursuit of their interests.

While completely authentic wilderness areas may not exist, there is a case for suggesting that areas that provide a semblance of wilderness and areas that can be managed to offer people the wilderness experience they desire is what matters. However, Dudley (2011, p. 45) has argued that naturalness is not irrelevant and that 'natural ecosystems are not just social constructs… they have existence and value distinct from human society'; later in his book (ibid., p. 189) he raises a series of key questions about the negotiability of authenticity:

> how much authenticity we want or are prepared to accept and what 'sort' of authenticity? Do we want all aspects of the ecosystem to be authentic or just some elements? Is the composition of species more important than the ways in which they interact? Is it worth restoring an original mix of species even if it risks upsetting the current mix or reduces the total number of species in the ecosystem? Do we need to trade off the natural regeneration processes against the social and economic upset that they cause? Does it matter that a few new species have moved in to replace the old?

He concludes by commenting that 'the answers will depend on time, place and what people want, but now we don't usually even ask these questions' and he stresses that such questions 'will become even more important as ecosystems vary more dramatically under climate change'.

9 Squaring the circle of authenticity in film and photography

From mass-reproduction to the punctum

Introduction

This chapter explores the effect of the depiction of places in film and photography on place authenticity. The popularity of the camera has been linked inextricably with the growth of tourism (Sontag, 1977; Chambers and Rakić, 2012) and photography is one of the integral ways in which people process places and practice 'everyday aestheticisation' (Sontag, 1977, p. 109). Urry's (2002) notion of the tourist gaze was predicated on the concept that we consume places as picturesque, scenic, 'framed worlds' (Robinson and Picard, 2009), visible through the window and lens as we travel by. It has been argued that tourists pursue and reproduce mass-media representations of places when they go on holiday, in a 'circle of representation' (Jenkins, 2003) or 'hermeneutic circle' (Urry, 2002; Haldrup and Larsen, 2012) of archetypal palm trees on empty sandy beaches. This chapter extends the notion of the circle, exploring how some places are adapted to resemble the images that commodify them, using examples of different types of reality. The book suggests that some places are designed to be panoramic, made for flâneurs and spectacle, and are experienced and authenticated in embodied and imaginative ways. Squaring the circle of authenticity also involves considering the filtered, edited, augmented digital afterlife of places. Perhaps, in fact, images of places exist in a strong bubble, rather than a hermeneutic circle of reality, which is malleable, but ultimately constricting.

The chapter unfolds the different ways in which photography and film operate on both panoramic and micro levels in places, not simply by adding characterisation, but also elements of enchantment and magic (Robinson and Picard, 2009). Film locations provide what Beeton (2015, p. 114) describes as 'mythscapes' generating story worlds and inspiring daydreamscapes, which allow us to defy the dimensions onscreen; film tourists cover great distances, travel to parallel galaxies and collapse and telescope time, without ever needing to leave home. Virtual reality headsets have replaced '3D' stereoscopes, Pokemon Go! turns the world into a playscape. Despite the acceleration of technology, tourists are increasingly fluent in the language of film; Pallant (2013, p. 183) observes that we still recognise depictions of place whether the backdrop is hand-drawn, computer generated or stop motion. When spaces are used as film locations, enhanced and filtered by cinematography, myths are invented (Hobsbawm and Ranger, 1983) and material places become imaginaries such as Clarke's (2005) 'cinematic cities'. Paradoxically, as photography and film strive to encapsulate the essence of places, the enhanced onscreen locations that they depict become fleetingly more auratic than the plainer tones of reality.

The chapter first discusses notions of mass-reproduction, an understanding of which is key to any discussion of filmed and photographed places. The chapter then moves on to film location tourism, where multiple realities are at work: for example, places substituting as others on film shoots or being romanticised or memorialised by their status as film locations. The section is divided into an examination of the nostalgic places of heritage film tourism, and an exploration of how fantasy locations are developed as forms of pseudo-history, reimagined by film tourists in the real world. The chapter then suggests that authenticity can be imagineered, giving examples of how some film locations have used mixed reality to become movie attractions. The chapter goes on to examine photography, place and tourism, discussing how the performance of authenticity is apparent in the staged and unstaged composition of images. Conceptually, authenticity can be linked to time and anachronism; the section explores how authentication emerges in the 'pauses' or detail within photographs, showing how a deeper chord of authenticity is struck by the 'punctum' (Barthes, 1980). Finally, the chapter investigates photographic placemaking in the more-than-human world.

Perhaps more than any other, this chapter establishes how copies of places fit inside each other self-referentially like nested Russian (Matryoshka) Dolls' (the "Matryoshka design principle"). The authentic inmost place is always concealed a layer away.

Representation and seriality

An understanding of different theories about the mass-reproduction of images is key to discussions of place authenticity, photography and film in the age of memes, sequels, reboots and retreads. Benjamin (1936 [2008]) described the 'demystification' of aura as part of his discussion of the visual reproduction of artworks in the age of film and photography. He argued that artworks are associated with ritual and high value. Mass reproductions circulate in the realm of the everyday, available to everyone, depreciating the value of the original; repetition 'severs them from their places in the fabric of tradition and thus erodes the aura of originality, uniqueness and permanence' (Frank, 1989, p. 29, cited in Steiner, 1999, p. 87). Benjamin was writing in the age of early modernism, when photography and mass printing revolutionised how people could see works of art. As discussed in the introduction to this chapter, authenticity folds in on itself in the mass circulation of cinematic and social media images, which is said to be a 'hermeneutic circle' (Jenkins, 2003; Urry, 2002; Haldrup and Larsen, 2012), a bubble, where the iconography seen on websites or in brochures, guide books and magazines is unconsciously sought and reproduced by tourists. Kodak Picture Spots in Disney theme parks provide places to stand to capture the optimum image, standardising the photographs people take and recycle. Roesch (2009, p. 129) uses the example of 'sunset point' at Uluru to explore how people strive to create idealised images: 'Representational practices initially created this vantage point, which was established as the perfect point to gaze upon, frame and photograph this famous monolith.' This practice of appropriation resembles the 'tourism sacralisation process' (MacCannell, 1989) involving the 'naming, framing and elevation, enshrinement and reproduction' of attractions, each phase unfolding another layer of constructed authenticity.

Seriality, or the creation of repeated images, is both an exact art and an impossible science, because often the aura of the subject is elusive. For example, Georgia O'Keeffe captures and recaptures New Mexico by painting serial Southwestern motifs such as skulls against a sky. Her repetitions of her patio door never quite evoked the reality and the door became an obsession, much like Churchill's pond (Barson, 2016, p. 15). Monet painted and repainted over 30 versions of Rouen cathedral in different light at different times of day 'creating a story of the passing of time' (Canjels, 2011, p. X1v). Relph (1976 [2008], p. 57, citing Cassirer, 1970, pp. 160–1) refers to Richter,

> who set out with his friends to paint the identical landscape in Tuscany while staying as close to reality as possible: 'Nevertheless the result was four totally different pictures, as different from one another as the personalities of the artists.'

Scenes can be reproduced on postcards, calendars, prints, newspapers, the internet, places can become memes, but Benjamin (1936 [2008]) attests that the aura is never truly replicated.

Benjamin's (ibid., p. 40) contention is that when objects are mass-reproduced (as images) aura is *replaced* by signs of authenticity: 'nevertheless, the function of the concept of authenticity in the contemplation of art remains unambiguous: with the secularisation of art, authenticity supplants cultic value'. He argues that people try to de- or re-mystify objects by authenticating them, which, in the contemporary world, would include photographing and filming places. Cohen (2012) suggests that marking or framing objects as authentic and removed from the 'flow of life' makes them cease to be authentic. However, Larsen (2008, p. 145) contends that contemporary digital camera technology and downloads create virtual flows that can be added to other forms of spatial and creative flows, which are 'man-made *and* machine-made'; photographers are hybrids: 'networked-camera-tourist'. Social networking is designed to involve other, absent people in photography, a networked-camera-tourist-society; the image is intended to be displayed to these absent others. Images themselves are fluid, reflecting 'imaginative geographies' (De Certeau, 1984, p. 105) and flows of place, as Urry and Larsen (2011, p. 167) state: 'Due to rapidly travelling images, places are, in effect, on the move, connected to other places and consumable at a distance.'

Benjamin (1936 [2008], p. 40) implied that mass-reproduction makes the aura of the original object or building increasingly vague. 'The more the cultic value of the image is secularised, the less specific ideas of the substratum of its uniqueness become.' Benjamin contends that trying to bring the auratic closer through mass-reproduction is a symptom of modernity and in doing so, the idea of 'genuineness' or 'historical witness' begins to 'wobble' and aura 'shrinks' and 'fades' (ibid., p. 7). Encountering the Parthenon, gazing on the real structure after having seen so many reproductions, Freud felt a sense of unreality (Rojek, 1997, p. 56). This example suggests the sensation of hyperreality, where copies have replaced objects and where places in films or photographs feel more tangible than actual locations, which instead become symbols. Unsurprisingly, the reaction is then to photograph the original and create another reproduction as a piece of real evidence of the viewing.

Benjamin's theory is still highly relevant; however, mass-reproduction and virtual authenticity are ever-sharpening our holistic, 360-degree view of places as is our use of devices as progressively organic memory prosthetics. The concept of aura is evolving. Rather than an authenticity turn, perhaps we should argue that authenticity should be turned on its head: the ubiquity of photography is such that a post-truth, or post-post-authenticity tipping point has been reached. More than ever, places exist to be reproduced, the quantity of mass reproduction is evidence of originality, and we seek to make these reproductions ever more lifelike in a cycle of 'real to fake to virtually real' (see Figure 1.3).

Film locations

Film location tourists are searching for places that resonate with aspects of the completed movie and television programme atmosphere. It has been suggested that authentic film tourism experiences are existential (Frost and Laing, 2015) and emotional (Beeton, 2005, Roesch, 2009); they are certainly experiential and offer an element of backstage insight (Beeton, 2015), which also makes them cognitive. Films create powerful associations with places as Beeton (2010, p. 114) states:

> films not only present places as a backdrop to the action, but also, at times, as characters of the narrative, which provide a level of emotional attachment for the viewer, especially when the setting is integral to the story and, in effect, a star in its own right.

Settings are essential for films; in a more-than-human world evoking mood and tension, shaping plots and generating dynamics. They also leave something to the imagination, to be inferred from what is depicted, from glances offscreen and the hints of the unseen (Lefebvre, 2006, p. 1). Heron (2004, p. 61) discusses the 'film landscape', stating the importance of 'establishing shots, informing the viewer of the role a place is likely to have in film'. An example is the opening sequence of the television series *Shetland* (2013), where the sky and vast landscape add mood, atmosphere and scope. Film locations seem to enhance aura and mystery for the romantic gaze as we project qualities onto the narrative to find symbolic meaning in them (Lefebvre, 2006, p. 48). Heron identifies three different types of roles that place plays in film: 'Landscape processes and forces are not important; the land dominates relationships, the land dominates the story' (Heron, 2004, p. 61). Where the land dominates the story and relationships, it can play a substantial role as a character in supporting or protagonist roles: 'landscape reinforces the tone of the human characters. For example, a dark, brooding land may be used to support an evil character' or 'a landscape that claims its own victims (causing death through exposure to the elements)' (Beeton, 2010, p. 116). The landform and weather of Iceland, in the television series *Trapped* (2015), involves blizzards, which control the plot by reducing the characters' range of movement, changing the landscape from scenic to hostile. In the same way, Monument Valley, which is used by John Ford in movies such as *The Searchers* (1956), is more than a backdrop to the action, it becomes a force: '[a] harmonious, immutable and gigantic setting, capable of enclosing the quest of the Fordian heroes in the midst of gigantic murals and pillars that unite the earth and the sky' (Maudet and Henriet, cited in Lefebvre, 2006, p. 48).

Heritage film tourism

Nostalgia plays a strong role in the popularity of film locations, as does the process of retrochic (Samuel, 1994). *Stately Attraction Film Tourism*, a report to the BFI by Olsbery/SPI (Jones and Kirby, 2007), examines the link between tourism and television. The report observed that people do not visit places simply used as action backdrops: 'In general, tourists are attracted to sites which are strongly associated with the story. Where a film/programme features a strong, real, historical character, tourists visit places associated with the real character, as opposed to the locations used for filming' (ibid., p. 10). Heritage film tourism selectivity is also apparent; the 'aura' of many places is vested in the authentic connection to the past, but not to any past – to specific, idealised mythical eras of the past, such as the Tudor or Roman periods (Lovell, 2010). Nostalgic identity was fortified through British Second World War propaganda films such as *Mrs Miniver* (1942), setting the scene for a white-cliffs Britain, establishing tropes of home-place and fear of loss. These films were of their time, but constant repeats on television continue to perpetuate place myths. The report *Stately Attraction Film Tourism* discussed the benefit of historical films and programmes which reinforced the identity of the UK as a historic country and provoked interest in seeing other places related to the era of television series such as *Downton Abbey* (2010–15). 'Gosford Park appears to have sparked a renewed interest in Victorian and Edwardian era historic houses, particularly those with a focus on servant's quarters and life "below stairs"' (ibid., p. 14). *Outlander* (2014), partly filmed in Scotland, reinforces the romance of the Jacobite uprising and eighteenth-century Scottishness and heritage tropes such as castles, aimed, in particular, at the American market.

This distillation of nostalgia into real-fake cultural stereotypes also applies to American Westerns, which are underlaid with the romance of the pioneer era in America, captured in *Little House on the Prairie*. The *Silverado* set is still maintained on Cook Ranch near Santa Fe in the Sonoma desert in New Mexico. Bonanza Creek Ranch operates as a Wild West movie set used for films such as *3:10 to Yuma* (2007). In New Mexico the scenery, including 'jagged mountains and piercing sky', is celebrated, but there is also 'a peculiar sense of dread that this gorgeous open territory may soon be swept up by the forces currently diminishing open space in the American West' (ibid., p. 100). The Western director John Ford was working in the 1950s, during the time of the largest expansion of the American Interstate Highway System, new middle-class access to cars and the growth of the countercultural movement (Laderman, 1996, p. 41; Beeton, 2015, p. 142). This was the era when the notion of Manifest Destiny inherent in the pioneer's journey westward was appropriated by the countercultural movement, the Beat poets crossing the land in the railroad cars, travelling as 'bums' and hobos, finding deeper, authentic existential selves in jazz, as detailed in the 'rugged white male individualism' of *On the Road* (Laderman, 1996, p. 41). Road trips celebrating freedom of movement across the West were celebrated in the 70s in films such as *Easy Rider*, *Badlands* and *Duel*, then in the 1990s in *Wild at Heart*, *Kalifornia*, *True Romance*, *National Lampoon's Vacation* and *Thelma and Louise*. Laderman (ibid., p. 42) suggests that the films also demonstrate how movies were 'progressive and politicised' in the 1970s, but 'depoliticised and socially indifferent' in the 1990s, thereby changing the framing of the original pioneer's journey westwards. In each of these films, the journey contains challenges, murder and death, suggesting the hostility of vast space, where travelling authentically and untrammelled is highly visible and any expressions of counterculture and notions of adventure are punishable by more conventional forces. The West is empty and transcendental, but full of diffuse danger.

Onscreen places often depict authentic normality. The ordinary world is portrayed in television soap operas and approached more as popular culture than art form. Seriality is viewed as a mass-reproduction technique (Canjels, 2011) and television serials allow the unfolding familiarity of places and encroaching affection, as different elements are revisited and rediscovered each week. *Dad's Army* (1968–77) displays multiple levels of 'nostalgic film tourism' (Beeton, 2005, p. 10), creation and re-enactment, demonstrating hierarchies of authentic (filmic) associations; for example, Cullen (2016) details how *Dad's Army* re-enactors are disapproved of by home guard re-enactors because they don't depict the 'real' situations. The battle for Walmington-on-Sea, the fictional town in which the series is set, illustrates layers of film authenticity. The outdoor scenes of the original TV programme were actually filmed in Thetford in Norfolk, where a statue to Captain Mainwaring was erected in 2016 to commemorate the opening of the film, complementing the *Dad's Army* Museum, which exhibits memorabilia contributed by the *Dad's Army* Appreciation Society. Nearby, the Bressingham Steam Collection houses a full-scale, faithful re-creation of Walmington-on-Sea, including Jones' butcher's shop, Fraser's Funeral Parlour and the Swallow Bank. In the 2015 film version of *Dad's Army*, Bridlington and Beverly in East Yorkshire were transformed into the fictional Walmington-on-Sea in the 1940s (Murfitt, 2015). Within the TV series, references to Hastings indicate that, if Walmington-on-Sea had existed, it would be located on the south coast, possibly near Bexhill-on-Sea.

Nostalgia is deepened by the threat of danger outside and the response within a close-knit community. Films such as the Aardman's *The Wrong Trousers* (1993) and *A Close Shave* (1995) are a study in ordinariness, a stop-motion nostalgic 1950s Yorkshire of small terraced houses, bakeries, Wendsleydale cheese and knitting shops, which is infiltrated by the criminal bank robber penguin and baker serial killer. The threat within is encapsulated by the Netflix television series *Stranger Things* (2016), set in small-town Indiana in 1983 and based heavily on the novels of Stephen King and the nostalgic films of Steven Spielberg, such as *ET* (1982). Small-town USA is generically familiar from *Breaking Away* (1979), *Twilight* (2008), *Fargo* (1996), *Peggy Sue Got Married* (1986), *It's A Wonderful Life* (1946), *Back to the Future* (1985) and *The Vampire Diaries* (2009). There are the usual tropes, such as Main Street, a café where the kids hang out, a bar, a library, a High School. *Stranger Things* (2016) inverts the tradition by presenting a post-apocalyptic view of small-town America; dark, abandoned, coated with organic growths and what looks like nuclear fallout ash. As with *Twin Peaks* (1990–1), unsettling strangeness and horror is most clearly offset by a backdrop of normality. The 1970s in America experienced a rash of films which unsettled ideas of normal place, for example the *Deer Hunter*'s (1978) hunting scenes in the Allegheny Mountains starkly contrasted with the scenes in Vietnam, the wider world from which the small-town Clairton community cannot be protected. Movies such as *Deliverance* (1972) and *The Wicker Man* (1973) depicted the rural communities as sinister and *Jaws* (1970) revealed the threat in the resort waters. Horror films such as *Friday the 13th* (1980) accentuated the fear of the woods through the terrorisation of summer camps, playing into campfire ghost stories. These examples demonstrate how film resituates places, by portraying certain essential aspects, changing the perceptions of people trying to experience the 'real' film locations outside the bubble.

Fantasy

Films and photography mix historical authenticity and enchantment in what Fjellman (1992, pp. 60–2) termed 'Disney realism' or 'distory', explaining how 'imagineers' 'improved' well-known narratives to the point of 'historicide'; adding to which, Massey (2013) has detailed the idea of 'global sense of place', which indicates the presence of a meta-mediascape, potentially diluting ideas for the mass market 'screen consciousness' (adapted from Olick, 2003, p. 8). Distory uses poetic licence to create a meta-narrative of history, distilling it and condensing it into an ahistorical, corporate package of global 'metafictions' (Hollinshead, 1998, p. 58). This can potentially provoke the invention of 'fakelore' (Crang, 1996), or 'bending history' (Hewison, 1987) to entertain and amuse. As tastes change, traditions are invented (Bruner, 1994; Hobsbawn and Ranger, 1983, p. 1), mutating the authentic experience into the 'experience economy', or 'the past as a plaything of the present' (Samuel, 1994, p. 429). A number of fantasy dramas are ahistorical – for example, *Game of Thrones*, which uses the motifs of pseudo-medieval places.

Animated places incorporate hyperreality within landscapes, such as the house ballooning away in *Up* (2009); 3D films enable other aspects of places to be revealed – for example, the surreal landscape of *Avatar* (2009) (Pallant, 2013, pp. 193–5). Fantasy films are popular because of spectacle, technology and nature (Pheasant-Kelly, 2014, p. 122) and their landscapes lend themselves to off-location tourism – for example, studios or theme parks such as Harry Potter World, with exaggerated, immersive experiences. Films may shape the perceptions of real-world places, as detailed by Gravari-Barbas and Graburn (2016, p. 12): 'Japanese animated films, embracing much of the wider fields of Japanese popular culture, are a major source of imaginaries for young Europeans, in creating a semi-fictional and fantastical Japan.' Fans of Japanese amine movies, such as Miyazaki's *Spirited Away* (2001) can visit the 'real-time' locations of the films (Jinto, n.d.), which presents specific challenges, as detailed in the following section.

In contrast to ordinariness, fantasy superheats magical realism. Puzzlewood, in the Forest of Dean, has misshapen, twisted pathways, rock formations and trees that inspire the projection of magical qualities. It has been used in *Merlin* (2008), *Dr Who* (2009) and *Star Wars: The Force Awakens* (2015) because it seems otherworldly and ancient. Roesch (2009) has researched how film tourists behave in fantasy locations such as those from *Lord of the Rings* – for example, in choreographed guided tours of New Zealand (Edensor, 2001). Roesch (2009, p. 129) observes that 'the romantic gazer' is solitary, possesses sustained immersion and the gaze involves 'a sense of awe'. Film tourists re-enact scenes, performing in place, using dialogue and movement and recording their behaviour, posing for or staging photographs. They recreate shots, displaying 'behavioural insiderness' (Roesch, 2009, p. 159; Relph 1976 [2008]), some taking the behaviour a stage further – seeing the scene and place through the eyes of the character, consuming an imaginary place in the real location. Lovell's (2013) 'mechanistic authentication' is apparent in their interaction with site markers, such as signs, and the memorabilia that they bring to the location to be photographed in place (Roesch, 2009, p. 205).

The actual encounter with a filmed place can be intensely spiritual and reverential. Fandom implies deeper levels of devotion and experiential authenticity (Wang, 1999), moviescapes can become 'landscapes of the mind, offering displaced representations of desires and values… [being] both material and mediated' (Harper and Rayner; 2010, p. 21). Beeton (2005) poses that, according to some film tourists, the film can seem more real than the place. The film tourist overlays the physical environment with the second dimension of filmed environment to form an imaginary space; when this is successful, the authenticity emerges and retreats (Lovell, 2013). According to a member of staff at the West Highland Museum, tourists stand at the side of the loch to take pictures of the position where Hogwarts should be, if its Harry Potter CGI incarnation existed in picturesque, but unmagical landscape.

Imagineered authenticity

Some film location tourists travel to see a precise place they have seen on screen. The location itself can be reduced and distilled to a focal point within the film, 'such a focal point can be a set piece, a plaque, a signpost or a specific view' (Roesch, 2009, p. 11) – for example, the Devil's Tower in Wyoming, location for *Close Encounters of the Third Kind.* A variety of knowing, falsified environments are involved in film-induced tourism; placemakers or 'imagineers' (Fjellman, 1992) encourage visitors by staging tourist focal points, merging fiction with local space in order to forge a mythscape. The blurring of fantasy and reality generates physically immersive fake-real experiencescapes by augmenting reality to incorporate elements of fantasy sites that are explicitly provided for performance. Places established to imitate fictional locations are first- or second-order simulacra, which could be defined as fake-fake-real as they are copying imagined, rather than heritage sources. The attempt of imagineers to be faithful to their source, however, lends them some authenticity. King's Cross Station opened a Platform 9 ¾ attraction, where people can be photographed seemingly diving through the wall holding onto a trolley, then purchase memorabilia at a Harry Potter shop. Icons, such as statues of fictional characters, have been added to various places to reflect the expectations of film audiences – for example, Sherlock Holmes in Baker Street, Paddington Bear at Paddington Station and Rocky in Philadelphia. Destinations changed to reflect fiction become attractions in themselves, absorbing the celebrity dust; for example, Carnforth Station, the location for scenes of *Brief Encounter*, has installed a heritage centre and refreshment room in 1940s style and the clock is a working clock featured in the film. Thus, places stay authentic to the fiction, 'true to the film', and can adopt an element of the aura of the original film. Another example is the Casa di Giulietta in Verona, owned by the Cappelletti family, which is alleged to be the inspiration for the Capulets in Romeo and Juliet. Tourists can stand on Juliet's balcony and wish for love by rubbing the breast of her statue. While these figures are fictional, there is a sense of fun about the consumption and their sense of aura is derived from and compounded by literature and film.

Fictionally augmented places are not always effective; Mexico City introduced a Day of the Dead procession drawing inspiration from the opening sequence of the James Bond film *Spectre* (2015), using giant skeletal marionettes. Some local people felt this cheapened the more intimate approach to the festival: 'They film James Bond here and now we have the "traditional Day of the Dead parade". Let's see what happens when [the mayor] finishes reading The Da Vinci Code' (Agren, 2016). A franchised fictionally real replica Cheers Bar in Faneil Hall in Boston recreates the original on which the show was based; however, large, plate-glass windows destroy the illusion of the cosy bar.

The 'mechanistic authentication' (Lovell, 2013) of film locations is apparent in the cooler comparisons made by tourists between real places and film settings, which creates issues for film location tourism. Sometimes places seem to inauthenticate the dream; complications can arise due to the negotiated reality of film settings: 'Sections might be added to existing buildings, entire structures constructed, or parts of the natural terrain changed, but the film would still be considered to be shot on location' (Konigsberg, 1997, cited in Roesch, 2009, p. 7). As Roesch (2009, p. 151) describes, some fantasy locations can seem less than their cinematic portrayal; 'spatial location discrepancies… distortions occur where imaginary places and real locations differ in their mutual recognition value'. He details the 'stretching or squashing of mountains' in *Lord of the Rings: The Return of the King* (2003) and refers to the example of 'creative geography' which amalgamates spatially separate places into one 'fictional destination'. The fantasy universe always contains unseen locations, including hints and clues about what is depicted outside the frame, suggested by glances offscreen (Lefebvre, 2006, p. 21). In *The Searchers* (1956), settlers have established a ranch and cattle in Monument Valley, although, in reality, it is a desert (ibid., p. 49). As Fox and Lovell (2015) detail, structured absence can mean that reality is altered to suit the film – for example, in *Breaking Bad* (2008–13) the Navajo are edited out of the desert scenes. Rosech (2009, p. 205) observes how some film location tourists are disillusioned, observing markers drawing their attention to the processes of tourism and seeing the artificiality or poor maintenance of sets. This lends itself to existential outsiderness (Relph, 1976 [2008]) where tourists feel alienated from their surroundings and are unable to suspend their disbelief and evoke their imagination. Tours of film locations, such as *Sex and the City* (1998) or *Gossip Girl* (2007) visit key sites for filming, such as the Central Perk building from *Friends*; Beeton (2005, p. 111) refers to the way in which film tour guides draw attention to the tour's exterior-only attractions, while interior filming occurs in studios. Beeton makes the point that, although this draws attention to the artificiality, some tourists go home with backstage knowledge about how films were made.

Places are disguised, dressed and CGIed due to financial incentives, the availability of scenery and conserved cityscapes and scenery; many films substitute one place for another in 'runaway' productions (Beeton, 2015). Visitors may find this disillusioning; for example, there is an expectation of seeing Venetian architecture in Cephalonia after viewing *Captain Corelli's Mandolin* (2001) (Hudson and Ritchie, 2006, p. 266). For some films landscapes become a convention – for example, Westerns must be filmed in the American West (Roesch, 2009, p. 64) because the repeated use of Monument Valley by the director John Ford in the 1950s made it a cowboy signature. The US Southwest setting of Western movies seemed non-negotiable; however, Westerns have increasingly been shot in Canada, Montana and South Dakota, represented by parks in the state of Alberta in Canada in *The Revenant* (2015); other movies such as *Brokeback Mountain* (2005) and *Unforgiven* (1992) were also filmed in Canada due to the financial incentives and scenery (Beeton, 2015, p. 114). *Star Wars IV* (1977) was considered to use the Western format applied in space; television series such as *Ripper Street*, filmed in Dublin, applies the Western feel to Victorian London.

During 2014–16 a new genre of period royal dramas focused on the royal courts of the UK and France in the eighteenth, nineteenth and twentieth centuries.

The authenticity was vested in detailed recreations of courts at a time and place. Historically intact streets, palaces and a monastery in Prague were used to represent eighteenth-century Paris in the street scenes in the TV shows *The Musketeers* (2014–16) and series two of *Outlander* (2015). The effect is constructed Paris, made of paint and plaster, façades which chip in the winter weather. The series *Versailles* (2016), filming at the same time, but recreating a slightly earlier era of the eighteenth century, was filmed on location in and around Paris. The television production of *Victoria* (2016) was shot using a series of stately homes in Yorkshire, with a 100,000 square foot set replica of Buckingham Palace built at Fenton, outside Tadcaster (Burrell, 2016), while the series *The Crown* (2016) used detailed replicas, standing sets of Buckingham Palace and 10 Downing Street built at Elstree Studios. In *Victoria*, Beverley Minster doubled as Westminster Abbey, whereas in *The Crown* Westminster Abbey was played by Ely Cathedral. The productions generated parallel worlds of royal unrealities, none of which were the real thing.

In contrast, the series *Game of Thrones* was known for its international natural locations representing the different kingdoms of the story. It used the Höfðabrekkuheiði, the Myrdalsjokull glacier and the Vatnajökull glacier in Iceland to depict North of the Wall in series two. Sensitive natural landscapes and World Heritage Sites can be affected by filming – for example, the World Heritage Site of Skellig Michael, with its surreal, beehive cells, a place where reality resembles a simulacrum, was allegedly damaged during the filming of *Star Wars VII* (Siggins, 2016). There were also problems because the shooting schedule coincided with the seabird breeding season – suggesting that the process of filming has implications for 'real' natural or wilderness places by transforming them into destinations, changing their character and associations.

Photography

This section of the chapter discusses photography, revealing different aspects of authenticity formed by bubbles of multiple realities. Photographs are objects, which are subject to authentication, and also portray subjects which can be staged, suggesting the intricacies involved in an already complex research topic. Authenticity has always been a key issue in photographic studies, most apparently in the division between 'documents and pictures' (Edwards, 2006). Edwards (ibid., p.12) cites Sekula's (1984) statement that: 'There are two myths about photography: the old myth that it tells the truth, the new myth that it doesn't.' Postmodernism implies that subjective interpretations of photographs are an acceptable, inherent part of the process and, as Edwards (2008) states, documentary photography contains an element of the staged. Both myths have been blurred further by fake news and, in the age of post-truth, Larsen's (2008, p. 144) views are prescient, as is evident in this comment: 'press photography loses its truth status and yet becomes more effective at deceiving us'.

Sontag (1977, p. 9) observed that photography develops 'in tandem with one of the most characteristic of modern activities: tourism' and is now a part of the way in which we process places. As described in the introduction, tourists have been said to be pre-programmed to pursue place myths and iconography as part of the bubble, or 'hermeneutic circle' (Urry, 2002; Jenkins, 2003; Haldrup and Larsen, 2012) of key images of destinations, gleaned from magazines and mass media, which they then 'track down' to reproduce. This involves creating an unreal, idealised version of place through photographic composition (Urry and Larsen, 2011, p. 169). Framing places in this way complies with the 'aestheticizing tendency of photography' (Crouch and Lübbren, 2003; Sontag, 1977; Urry and Larsen, 2011) which can be particularly powerful; images no longer present a 'window on reality' (Albers and James, 1988, p. 150, in Jenkins, 2003, p. 314), but offer a 'new reality' (de Clippel, cited by Canal in Alfonso *et al.*, 2004, p. 33) where the images mimic or simulate the world in heterotopias, forming parallel worlds which mirror our own. New apps such as Dreamscope add elements of magical realism by adding effects to places – for example, Van Goghian swirls, transforming places into fantastiscapes.

Staged and unstaged

The discussion about authenticity and photography is ongoing; *Street and Studio* was the subject of a 2008 Tate Modern photographic exhibition. The exhibition divided photography into the staged and unstaged; images which were posed in a studio and a more naturalistic paparazzi shot, catching sight of a person in place. Some tourists incorporate places into their poses, people mirroring landmarks for example leaning by the Tower of Pisa (Haldrup and Larsen, 2003).

As with most discussions of authenticity, binary examples such as staged and unstaged can be excavated further for their complexities. One of the most common ways in which photographs are staged, as observed by Lovell (2013) in a visitor employed study (VEP) of 16 historic cities, was that some research participants attempted to capture places without other people in them. Other tourists are frequently edited out of tourist photography, which complies with a desire to conform to the mass-media bubble. The impulse to alter places on film to erase undesirable or less aesthetic aspects implies the concept of 'structuring absence', which was discussed earlier in the chapter. Barthes (1980, p. 42) describes these shadowy Others as 'co-presence', a duality created by the daily life of places, which continues despite the photographic composition. Other tourists can make scenes seem less authentic by reminding the photographer that they too are a transient visitor, undermining the experiential sense of authentic self while they travel. Barthes details the way that occasionally subjects who 'happen to be passing' can become the subject of photographs. Some tourists use photography like paparazzi, snapping a brief semiotic skimming of the city and the actors within it. The paparazzi street gaze is 'casual' and more of a snapshot, point-and-shoot, grasp, glance or blink, (Jenkins, 2003; Urry, 2002; Crouch, 2002, Henkel, 2014). This is photography not as 'a mirror of reality, but as ways of grasping the world', the 'theatre of life', which is arguably more authentic (Haldrup and Larsen, 2003, p. 25).

Rose (2001, p. 42) has observed that people trying to record the busy *mis-en-scène* (sometimes including other tourists) and what Jacobs (1961 [2000], p. 60) described as the metaphor of the 'sidewalk ballet' and the dance going on upon it are a key part of place. Many participants in Lovell's study captured other people as part of the overall atmosphere of place, occasionally observing tourist crowds. People photograph and post pictures of other people, in acts of flâneurie and memory-making, 'the activity now emblematic of the tourist: the democratised taking of photographs – of being seen and recorded and seeing others and recording them' (Jokinen and Veijola, 1997, p. 25). Sontag states that 'the flâneur finds the world "picturesque"' (Sontag, 1977, p. 55); flâneurie thus Others people, who become commodities or 'living advertisements' (Céline, 1988, cited in Clarke, 2003, p. 89) in the romances which are inscribed on the scene. Place is often reduced to a backdrop or prop for performances, which loses its meaning in the volume of pictures online, seemingly artificial, like a studio green screen set.

There is a similarity in the way that people re-enact scenes from films in place and perform stereotypical poses when being photographed. In their study of family photography at Bornholm Hammershus, Haldrup and Larsen (2003) found the romantic, thirteenth-century castle was positioned in some photographs to offset the aesthetics of the family on holiday. Other participants didn't include it because the photographs were designed to show a happy family at leisure. While *not* photographing the castle may seem to escape the hermeneutic circle of reproducing tourist imagery, photographing the happy family feeds into a different form of stereotypical images and poses.

Pause and flow

Rather than an artificially performed visual practice, photography could be considered as what Haldrup and Larsen (2003) and Urry and Larsen (2012) suggest is an organic part of the process of memory-making and performance. In a study of photography in museums, Henkel (2014) submits that taking photographs of each artefact impairs the number of artefacts subjects could recall, but zooming in on detail did not affect the powers of recall. Photographs are objects of memory (Barthes, 1980; Haldrup and Larsen, 2012) and, as Larsen (2005, p. 39) posits, they enable the tourist to travel back in time; the photograph is created as an artefact with future memories in mind. It could be said that photographs freeze-frame the past, providing 'artefacts of memory' (Crang, 1997) and capturing a moment when time and place are on pause, as demonstrated in research conducted by Lovell (2013) where a photograph of Lincoln Cathedral contained the caption, 'Time stands still'. The tourist photograph could be argued to offer the objective authentication of being taken in real time, but portraying unreal time and sometimes preserving places that change or disappear in the future.

Photography authenticates the experience of being in a place, photographs are haptic souvenirs (Sontag, 1977); evidence of being 'right then, right there' (Scifo 2005, cited in Larsen, 2008, p. 149) and offer a sense of the 'immediate experience' of place (Relph, 1976 [2008], p. 53). However, black and white and tintype photographs reveal a sepia past that colours our images of places as more distant from us; even if they are tinted, this is apparent as a modification, also changing the way we view places in earlier times. The increasing popularity of tintypes and box camera processes is a sign of retrochic (Samuel, 1994) and pleasure in the mechanistic authentication of anachronistic practices that were unpicked further in Chapter 4.

Perhaps one of the most eloquent ways in which photographs authenticate place is through the backstories that they can provide. The viewer may skim the image when reading it, in the same way that a photographer may skim the text of a place. Barthes defines the 'studium' (1980, pp. 20–8), a way of studying photographs, as the 'application to a thing, taste for someone, a kind of general, enthusiastic commitment'. Photographs can be affective, within a scene a single detail can stand out as poignant to the viewer, described by Barthes, (1980, pp. 26–7) as a 'punctum', which gives pause for thought. For example, he cites a bandage on a finger, which prevents the photograph being 'inert' under the gaze. In photographs of historic cities, Lovell (2013, pp. 216–17) found details of the punctum to include 'a lamppost with a wonky top' and 'a sculpture of a cat which a local architect in York places in the window of his buildings as a motif. The architect has died and the cats still appear to be waiting for him.'

The narrative offered by the punctum can allow the viewer to zoom in on a detail, draw out local memories and be personal and eccentric, providing a backstory for places. The punctum is also a way of finding the poignant in the everyday location. Larsen states how Goggin (2006, cited in Larsen, 2008, p. 149) describes the use of camera phones to illustrate our fleeting everyday observations, allowing people to capture the poetic in their daily lives.

Photographs can reveal multiple authenticities and perspectives when they are studied. Point of view theory (Rose, 2001) was first developed as a method of semiotic analysis, a tool which was used to investigate the cultural meanings of advertising and other mass-media images. Points of view are identified using 'mis en scene, including focus, angle, point of view, screen plane, superimpositions' (ibid., p. 42) and, in addition, poetics – enabling tourists to tell a story with their compositions. The meaning of photographs of places is constructed; the image therefore becomes a collaboration between the perceptions of the photographer and the viewer, 'both of whom carry their social positions and interests to the photographic act' (Harper, 1998, cited in Prosser, 1998). Photography betrays the 'the multiplexity of everyday life' (James *et al.*, 1997, cited in Alfonso *et al.*, 2004, p. 160) and its collaboration, negotiations and liaisons (Harper, 1998; Pink, 2001).

Photographed places are increasingly in a state of perpetual motion; the distinction between photography and film is progressively blurred as demonstrated by panoramic shots and Google's 360 degree views. Memories of hot or warm locational authentication travel with mobile pictures and site themselves where photographs are framed, pinned, or Instagrammed as emotional memoirs. Benjamin (1936 [2008], p. 6) observes how photography makes 'it possible for the original to come closer to the person taking it in… A cathedral quits its site to find a welcome in the studio of an art lover'. Images are fluid entities, illuminating 'imaginative geographies' (De Certeau, 1984, p. 105) and conveying the flow of presence.

Photo placemaking

Photographers such as Ansell Adams have articulated ideas about place in specific ways – for example, his black and white images of the American West, national parks, and the scenery of New Mexico and Yosemite – framing places as transcendental to impart an environmental message. Henri Cartier-Bresson's idealised photographs of post-war Paris underpinned its reputation as a romantic capital. Martin Parr, in contrast, contends that most photography is 'propaganda' and uses garish, grotesque and exaggerated portrayals to reveal our cultural peculiarities and those of tourist sites (Parr, n.d.). Ordinary people and their surroundings are given status as the subjects of his work, with evident humour at tourist behaviour.

It could be argued that tourists have a vested interest in conveying the romantic idyll of their travels, as stated by Robinson and Picard (2009, p. 8):

> the photogenic is that which invites artistic composition by virtue of its fit with preconceived notions of artistry to the point of the stereotypical with views being constructed and replicated based upon accumulated knowledge and culturally located ideas of what is attractive.

Ziller (1990, p. 37) suggests that, in the photographic process, subjects are predisposed to photograph the positive and, according to Sontag (1977), avoid the ugly. Photographs are heavily edited (in the same way that placemakers use selective marketing imagery), preserving memory highlights and deleting less entertaining aspects of the experience. Like films, photographs are not holistic; they are fragments – there is always something out of shot and incomplete (Kasfir, quoted in Kürti, 2004, p. 57). Compositions may disguise as much as they reveal – for example, a photograph showing a native hut can edit out a satellite dish on its roof (Markham, 1997, p. 132, cited in Jenkins, 2003, p. 319) and, in this way, photographs only ever provide what Pink (2001, p. 8) suggests is a 'semblance' of the whole place.

Habitual usage contributes to visual perception, developing 'the good eye', and practising a 'visual connoisseurship' (Rogoff, 1998, p. 17, cited in Rose, 2007, p. 35). It is argued that the act of photography could allow people to authenticate environments more strongly by looking through the lens (Garrod, 2008, p. 385) to visually 'grasp' (Haldrup and Larsen, 2003, p. 25) scenes. Investigating the links between tourism and heritage buildings, Willson and McIntosh (2007, p. 77) discuss the aesthetic nature of architectural tourist photography and how open public spaces provide material visual cues and clues, making them legible for photographic study (Lynch, 1960, cited in Emmison and Smith, 2000).

Photography provides a method of impulse-led authentication for tourists; there is a demonstration effect apparent in tour groups (Lovell, 2013, p. 241) – when one person takes a picture, others automatically follow. Lovell also details how the act of taking a photograph can mechanistically authenticate the subject; tourists may photograph places because of the facts surrounding them. As one tourist commented; 'It must be an important pub because everyone was taking a photo of it.' Pink (2007) describes how film and photography thus allow tourists to perform a placemaking activity of their own. Some photographs however are limited in the sense that, rather than 'being there', they inscribe placelessness or they describe 'being where' (Crouch and Lübbren, 2003, p. 6), with no visual markers that could identify the locality (Garrod, 2008).

Benjamin (1936 [2008]) discussed aura in the age of mechanical reproduction and the chapter has conveyed a glimpse of how 'places and their contents are seen from numerous angles and are apprehended as fractured but recomposed in mental processes' (Crouch, 2002, p. 212), as the afterlife of places ensures they are preserved for posterity in the collective memory of the cloud. The show *CSI: Cyber* frequently tried to solve crimes by accessing people's photographs on social media, compiling them into a large jigsaw of action caught at a specific point in place and time, a technique likely to become a social media inter-experience and, as Kwon (2004) observed, allowing places to 'assemble themselves'. This is not to say that there is no depth; collating myriad images of the eccentricities and micro-stories of places and the elusive backstage becomes increasingly transparent, but the sheer volume of images obscures reality and makes it harder to authenticate with veracity.

Conclusion

The chapter has squared the circles of representation by demonstrating how photography and film shape the composition and commodification of places, inspiring recreations and re-enactments. Photography is increasingly ubiquitous in the modern world; it serves to preserve places outside the memory, authenticating visits, but, while photographs are objects, their authenticity is negotiable and subject to multiple interpretations. People have always relied on prosthetic forms of memory and, rather than a visual practice, photography could be considered an organic part of the process of memory-making and argued to be an authentic, non-representational act. However, memories can change over time, making photographs unreliable narrators. Jenkins (2003, p. 319) states that 'photographs give people an imaginary possession of a past that is unreal, they also help people to take possession of space in which they are insecure'. Research by Henkel (2014) and Yagi and Pearce (2016) has indicated that where photography was not permitted, memories of places visited were unclear, suggesting that there may come a time when we cannot truly recall places which we have not photographed. The chapter explored how the authenticity in photographs is textured, going beyond categories such as staged and unstaged. Places are increasingly in disguise, suggesting that on the carapace of the social and material fabric rest many different imagined geographies; images can create and play with the fluidity and negotiability of reality on multiple levels.

One of the key questions of this book is whether the authenticity of places ceases to matter in a post-postmodern world, with its abstract, infinite cloud and virtual ways of consuming places continuously referring to and revealing each other like Russian dolls. Lovell's (2013) research on staged authenticity in historic cities suggests that the mechanistic authentication of places has become more important to visitors, who orientate themselves in fantasy environments and real worlds by pinpointing the origins of film locations. Film tourism exhibits combinations of the front and staged backstages of authenticity, including an imagined frontstage, where tourists overlay the story onto the scenery. One of the strengths of this text is that addressing different subjects allows the cross-fertilisation of ideas; the book reveals similarities between heritage re-enactors and movie fans. Both require accurate foundations; it is only when their two worlds are aligned that their fantasy performances can take place. Visiting film locations and trying to equal the peak sensations found through cinematic viewing in an immersive environment may prove far more difficult. Dissonance can occur when expectations are not met; however, some tourists enjoy aspects of mechanistic authentication when provided with the insider aspects of the film location transformation. The chapter examined how the hermeneutic bubble of tourist imagery constrains our self-expression and shapes our aesthetic appreciation of places; places are imagineered, as themed replicas of fictional places are realised; real places are commodified; nostalgic cinematic interpretations involve the reappraisal of places through emergent authenticity; and some real places seem to inauthenticate our fantasies of them. George RR Martin pays with the plasticity of historic events for *Game of Thrones*: 'I take it, I file off the serial numbers, then I turn it up to 11 and change the colour from red to purple, and I have a great incident for the books.'

The book has described how gentrified and themed places are intended for spectatorship; the growth of technology has continued to reinforce and extend the visuality and commodification of places. As indicated by other chapters, the authentication of places involves different phenomic perceptions and experiential states, including immersion, intellectual curiosity, irony and imagination. Global meta-fictions, films and images can alter the gaze as Donald (1999, p. 73) indicates, in a key observation about the animating and distancing effect of television, cinema and the mass media: 'The juxtaposition between the modern city and the technology of cinema provides more clues as to the way we experience the city as psychic space: the way we imagine the city.' Pallant (2013) discusses how augmented reality devices on tablets and phones using a combination of GPS and internet streaming overlaid on existing spaces provide future inter-reflections between image and place. The popularity of the Pokemon Go! experience confirms that the futurescape will be consumed on the move in a mixed reality of place, cartoon and filmed space, virtual, imagined and material. As the chapter suggested earlier, the aura of places, while elusive, is slowly being cemented by the sheer imperceptible heft of a cloud of images back into the frame.

10 Events and places

From giant spectaculars and carnival masks to hyperreality

Introduction

Events and festivals occur in many of the places that have already been covered in this book: in urban areas, seaside resorts, heritage sites and the countryside; in a variety of venues, from opera houses and pubs, to the streets and parks. Perhaps because they are temporary, perishable and therefore rare, events can play an important part in tourism activity and placemaking (Getz, 1997; Yeoman *et al.*, 2004; Bowdin *et al.*, 2006; Richards and Palmer, 2010). Events are an example of Third Space (Lefebvre, 1991; Soja, 1996), incorporating the physical, the designed and the lived experience of space. The varied range of events and festivals influences the transformation of the physical, intellectual and phenomic environment in terms of aesthetics, spectacle, atmospherics, colour, sounds and smells. Their transience creates heterotopias (Foucault, 1984), forming uniquely fluid and movable places overlaid on the physical environment and mingling the theatres of everyday life with adorned decorated scenery.

In this way, events provide a movable topology for the geography of authenticity and this chapter explores how events can authenticate places in a variety of different forms: by re-enacting the past, repeating rituals, banning tourists from an 'original' backstage and fostering a sense of local uniqueness. The chapter discusses ways in which events both articulate authenticity by bringing about a distinctive sense of place and mask places by generating consciously 'designed' experiences (Richards, 2014; Berridge, 2007). Events turn even the most natural of sites into explicitly staged and performed environments, front stages on which to 'gaze' (Urry, 2002) at cues such as lighting, ritual and costume. In the climate of the 'experience economy' (Getz and Page, 2010; Johansson and Kociatkiewicz, 2011; Pine and Gilmore, 1999; Robertson *et al.*, 2007) events may be regarded as 'real fakes' (Fjellman, 1992; Hannigan, 1998) which, as Lovell (2013) described when researching staged authenticity in historic cities, is where the staging is paramount.

According to Walter Benjamin (1936 [2008]), in *The Work of Art in the Age of Mechanical Reproduction*, the aura of places can be diluted by reproduction; this concept is one of the central ideas examined in this book. The replication of event formats is addressed by Yeoman *et al.* (2004) and Richards and Wilson (2006). An over-abundant supply of events may suggest both an increase in niche products and the development of events for the mass market. Increases in both leisure time and discretionary spending have led to a proliferation of public events, celebrations, entertainment and festivals, which have been identified as one of the fastest-growing forms of tourism-related phenomena, leading to what Richards (2010, p. 1) described as 'hypereventfulness' or 'hyperfestivity'. Richards (ibid.) quoted Boorstin's (1961) description of 'the programming of our experiences', which have 'no peaks and valleys, no surprises'. Boorstin also detailed the attributes of 'pseudo-events', which Richards goes on to summarise as 'encapsulating a lack of spontaneity (they are purposefully planned)' and with a strong media focus. 'Events have an ambiguous relation to the underlying reality of the situation, for example whether it is "real" or not is less important than its newsworthiness and ability to gain favourable attention.' This comment is highly relevant for events such as the 2016 US election, where the phrase 'post-truth' first gained traction in popular usage.

The chapter discusses the history of events, examining mega events, which could be described as 'spectacles', a term defined by Manning (1992, p. 292) as 'a large-scale extravagant cultural production that is replete with striking visual imagery and dramatic action that is watched by a mass audience'. In *The Society of the Spectacle*, Guy Debord (1983 [1987]) suggested that reality has been replaced by images and representations, the true intention of which is to mask capitalist commoditisation. To avoid their characterisation as 'placeless festivals' (MacLeod, 2006), many outdoor events are purposefully designed to add meaning to their location – for example, by embedding local distinctiveness, heritage and invented, or revived, traditions within the theming and creative process. The chapter assesses how some events can be firmly rooted in some peripheral communities that resist globalisation, de-commodifying their events by restricting tourist access to the places and venues where they are held. A further examination of community festivals argues that some events reflect the object of the gaze back to tourists in displays of staged authenticity and local customs, while retaining a private backstage. Different types of gatherings assessed in the chapter include carnivalesque environments and hyperreal spectaculars, which suspend reality for atmospheres of magical reality. The chapter then examines aspects of experiential authenticity, such as liminality, ritual, communitas and re-enactment. Finally, the chapter suggests that events can change places through their impact, providing a legacy for infrastructure and the potential for the construction of new identities.

History

Attributes of authenticity can sometimes seem to rely on the repetition of historic ritual and tradition grounded in simple longevity. Events have played a strong role in society for thousands of years, contributing to feelings of social and place attachment. Carnegie and McCabe (2008, p. 355) state that: 'There is a long tradition of literature in anthropology on the nature and meaning of indigenous festivals, rituals and celebrations which continues to resonate powerfully on the meanings and interpretations of enactments of cultural identity through the festival.' In the past, many events were purely local affairs, possibly involving no more than village communities coming together to mark important events in their lives or to celebrate the seasons or key dates in the religious or agricultural calendar, such as the Midsummer festival on 24 June in Sweden. Quinn (2005, in Finkel, 2010, p. 275) argues that events form and re-form local identities through 'ritually repeated actions'. As Finkel (2010, pp. 275–8) argues, 'what rituals people hold dear and how they entertain themselves can be informative for finding out about the way a society structures their belief systems'. Their abiding presence in the lives of people suggests the need to celebrate and commemorate, as if they are the physical manifestation of collective memory. As Berridge (2007, p. 166) states, 'Many of our lifestyle timelines are framed by specific events that leave an emotional impression upon us. They help us identify periods of our past and are used to give meaning to periods, people, places, occasions.' One of the earliest documented examples of such an event was the Olympic Games, which date from 776 BC. However, the Pan-Hellenic games at Olympia were but the most prestigious of more than 100 such festivals (Finley and Pleket, 1976). Athletic games were an essential part of Greek life and every self-respecting city had its own stadium (Davies, 1997), with the touring element being an important part of the sport. 'There was probably no other occasion in the ancient world when as many people were on the road (or on the sea) for the same destination at the same time' (Finley and Pleket, 1976, p. 53). Unlike provision for modern-day tourists, there was little accommodation and visitors slept in tents or in the open air, although a hostel was established at Olympia in the fourth century (Baker, 1982, quoted in Standeven and De Knop, 1999). Such an event must clearly have had a substantial effect both on the environment of Olympia for the duration of the games and on the longer established perception of the place as an event tourism destination.

Mega and hallmark events

Events range from small local events or festivals to what some have described as mega events, implemented for their regenerative capacity. Getz (1997, p. 6) defines mega events as those with more than 1 million visitors and capital costs exceeding $500 million, which 'yield high levels of tourism, media coverage, prestige or economic impact for the host community or destination'. Mega events affect the authenticity of places in several ways. They can amplify the rituals and cultural stereotypes of national identity through events such as Olympic opening ceremony spectacles – for example, in Rio a sweeping history was presented, ranging from re-enactments of the journeys of the First Nation to the growth of the local favelas.

The social and cultural impacts of events add a further dimension of authenticity – for example, the displacement of people to make way for infrastructure renewal changes the authenticity of the original sites. One of the defining characteristics of a special event or festival, according to Gilbert and Lizotte (1998, p. 73), is its 'transience'. Ferdinand and Shaw (2012, p. 12) citing van

Elderen (2007) support this theory, defining 'festivalisation' as 'the temporary transformation of place into symbolic space in which the public domain is claimed for particular forms of consumption'; however mega events strive to leave a lasting legacy focused on the community where they are staged. Like the gentrified environments discussed elsewhere in the book, the new mega-places are 'front regions' (Goffman, 1959, p. 110), shop windows for cities – for example, Rio's Port Zone, regenerated as a cultural showcase for the Olympics, featuring an art museum, the Museum of Tomorrow, Maua Square and a six-point trail of slavery sites in Little Africa.

Hallmark events are considered to be smaller in scale than mega events and generally have a special meaning or significance for a region or community (Cassidy, 2006). The event emerges, first, as an add-on, then as part of the fabric and essence, creating an event-place, in the cycle of 'fake to real' (Lovell, 2013). According to Bowdin *et al.* (2006, p. 17) hallmark events are 'events that become so identified with the spirit or ethos of a town, city or region that they become synonymous with the name of the place'. Thus, the event becomes a metonym for the place – for example, Glastonbury, Wimbledon or Rio. Il Palio, in Siena, is a hallmark event, an annual horse race in the central city piazza. It shows how events are inextricably connected to places on all levels, including political, ritualistic and traditional. It is the continuity of the staging – for example, the costume and the historic setting – which make the Palio seem authentically medieval, as do the local politics, rivalry and community negotiations that are an essential part of the pageant, as disclosed in the film *Palio* (2015). In the case of the Edinburgh Fringe, the city streets are taken over by the Festival to the extent of theming the place entirely, with multiple venues in use until an alternative Edinburgh, with its own event multiverse, is created. Richards (2015, p. 554) uses the Edinburgh Festival to apply actor-network theory to events: 'the festivals themselves also begin to take on active roles in shaping the idea of the city'.

Re-enactments and community events

Re-enactment events allow the recreation of historic moments, often in places such as battlefield sites, where there is no visible evidence of the past. Re-enactment events are not without their critics; Carnegie and McCabe (2008, p. 356) observe that some historians argue that re-enactment events are inauthentic:

> But we argue that the interaction between the place, performers and visitors allows for dynamic construction and interpretation of the way in which to interpret the event in terms of its authenticity and agree with Bruner (1994, p. 411) that living history 'can be a good way to learn about the past'.

They go on to quote Goodacre and Balwin (2002, p. 59, cited in Carnegie and McCabe, 2008, p. 352), who make the point that:

contact with the living past is impossible and those living history interpretations in museums or at historic sites are as much illusions as any other form of peopling of historic space. Their chief benefit is that they allow us to explore our relationship with the past through human contact.

They articulately make the case for the value of re-enactment and, as Crang (1996, p. 421) has observed, re-enactors are fanatical about detail, focusing more on the authenticity than the staging. The mass practise of authentication at a granular and lifestyle level would indicate that, far from re-enacting the past, the past is reanimating the present place (see Figure 10.1).

Event authenticity has been indicated to pivot on the relationship between hosts and consumers (Richards and Wilson, 2007; Simons, 2014; Finkel, 2010). Many community events involve festivals that are organised primarily for local people, with many demonstrating long-established ritualistic folk traditions that are fostered for their uniqueness and originality. Abbots Barton Horn Dance, in Staffordshire, features dancers wearing antler headdresses and characters, including a hobby horse and a fool. The Mari Lwyd, in South Wales, is a custom involving a hobby horse made of a horse's skull on a pole, which is used to accompany winter singing traditions. Community and folk festivals could be argued to be both the most rooted in

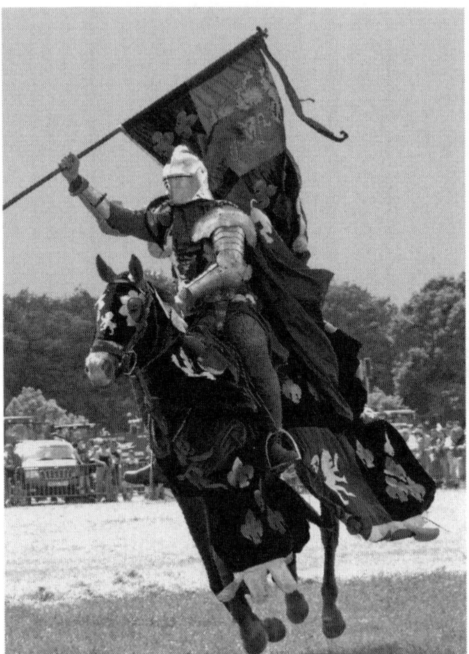

Figure 10.1 Knights of Middle England
Source: Knights of Middle England

place and, because of their distinctivesness, perhaps most vulnerable to commoditisation. Drawing on the example of the Draaksteken Beesel, the village of Beesel in the Netherlands that re-enacts the George and the Dragon fable, Simons (2014, p. 88) suggests that slow principles can be applied to event design, examining elements such as collective community decision-making. Simons (2014) makes the point that any changes to deeply rooted, slow events can affect participants' perceptions of authenticity, as if the rituals will be diminished by interference.

Many other community events are drawn from local cultural capital, re-enacting and reinscribing histories. Finkel (2010, p. 275) describes Up Helly-Aa, in Shetland (see Figure 10.2), an event involving the burning of a replica of a Viking longship. 'Through ritually repeated action that now translates as tradition, Up Helly-Aa interprets and reinterprets what it is to be a Shetlander. It is difficult to extricate this festival from place, community, local identity, and a specific world view'. Finkel (ibid., p. 282) goes on to detail how Up Helly-Aa is community funded, enabling local people to try to exclude visitors, creating a private, authentic space for communities. These private spaces enable the event to act as a catalyst for inter-generational and family stories and mythologies, as people weave their own narratives around their participation and reinforce their social networks and place attachment. Celebrations of Viking culture, including re-enactments, have been said to be subject to both commodification and the 'surveillance of authenticity' (Halewood and Hannam, 2001, p. 578; Hannam and Halewood, 2006). Finkel suggests that the event is less vulnerable to being Disneyfied by the tourist gaze because it is conducted in a peripheral community on Shetland, which is less susceptible to the flows of globalisation. Getz (2005, pp. 142–3) discusses how some residents choose to hold their events off-season because they seem more authentic due to a lower tourist presence. Banning tourists is a way of

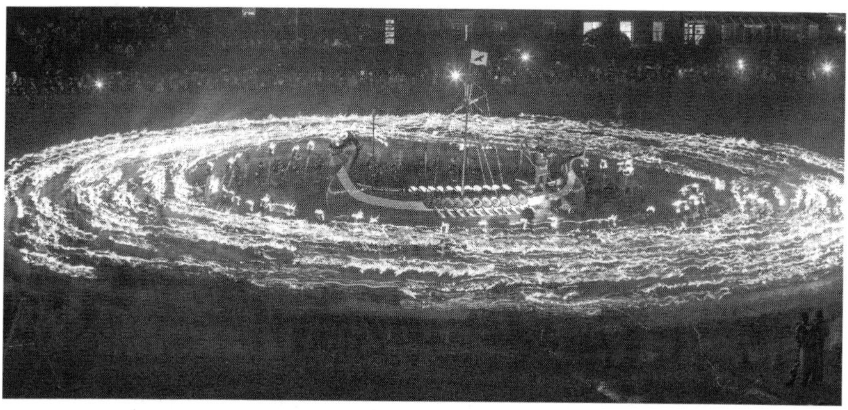

Figure 10.2 Up Helly-Aa encircling the galley
Source: David Gifford

de-commodifying the event; however, rarefying events in this way may simply increase their forbidden, backstage aura, attracting more tourists.

Urry (2002, p. 9) has described the process of staged authenticity, where event participants and 'local tourism entrepreneurs gradually come to construct backstages in a contrived and artificial manner' and some events show comparatively recent examples of new 'invented traditions' (Howbsbawn and Ranger, 1983) and reviving older rituals. In the process of connecting to international partners as part of a 2008 European Capital of Culture bid, Canterbury created Geants (giants) for their processions, as used in French towns (Lovell, 2009) and these figures featured in Canterbury processions. Such rituals can become established and absorbed into the spirit of the festival, is the recalibration of 'fake to real'. Relph (1976 [2008]) argues that nostalgically created places are inauthentic and, increasingly, today many such events commodifying heritage act as niche tourist products, both reinforcing local distinctiveness and attracting spectators from further afield. In this case, community events can provide temporary forms of gentrification, reinforcing the backstories of places and involving the local production of foods using traditional processes. For example, the Whitstable Oyster Festival celebrates the industrial heritage of the harbour through a variety of events such as an Oyster Trail and focuses on the past heyday of the place through the Landing of the Oysters at Long Beach, where the first catch of the season is blessed.

One of the aspects of events that may enhance or detract from their authenticity is their creative design. Events are intentionally created, in extreme cases by Disney 'imagineers' (Fjellman, 1992; Richards, 2014). Lovell (2013) has written about 'heritage opportunism' in historic cities and Bevolo (2014, pp. 74–5) argues that local character can be over-exploited: 'It is a common risk in fine arts and design that what was born as authentic, genuine, creative DNA becomes, over time, nothing more than a styling cliché, a marketing gimmick to cash in on.' For example, Bossen (2000) cites the case of Fiji, where various festivals, including the Hibiscus Festival in Suva, the Sugar Festival in Lautoka and the Bula Festival in Nadi, have become natural parts of the national culture. However, the festivals were started purely as attractions to lure tourists to Fiji. Such overtly staged authenticity has led to debates about its ethical nature and whether it is right to create partial or, as in the Fiji example, totally false representations of reality. Bevolo (ibid.) argues that 'The challenge will be to connect the power of design to the authenticity of vernacular values to prevent the risk of superficial imitation and meaningless repetition of vernacular events.' However, as Urry (2002, p. 151) has suggested, such developments may have the effect of protecting local communities by keeping out what may be deemed 'the intrusive eye' while providing visitors with what seems properly authenticated.

Hyperreality

Commemorative events are imbued with heritage, social ritual and collective memory; centenary remembrances are increasingly spectacular – for example, the

Blood Swept Lands and Seas of Red at the Tower of London in 2014 discussed in Chapter 5, which explores how some events are also designed as simulacra for a hyperreal effect and 'fairytale gaze' (Lovell, 2013). Hyperreal events can come in several forms – for example, rock concerts that use special effects, light shows employing projection mapping, mass events such as pilgrimages, surreal events such as art biennale, or events that play on sheer scale. An example of a third-order simulacrum is the Giant Spectacular in Liverpool, developed by the company Royale de Luxe who describe their style as more-than-human 'imaginary realism' (Moseley, 2012, p. 20). The giants are inspired by the folk legend that mountains are 'sleeping giants'. The company developed three 'time traveller' characters whose 'sheer scale would inspire the same sense of awe and wonder'. They play on fairytales; myths about giants have long been a central component of different folklore and the company reinforces this with invented mythology surrounding the giants, whose arrival has apparently moved tectonic plates. *Sea Odyssey* was performed in Liverpool in 2012. The event was staged to commemorate the centenary of the sinking of the *Titanic* and started with the arrival of a mysterious object, in this case an anchor in King's Dock. The parade involved the 9-metre high Little Giant Girl and the girl's Mexican dog, Xolo, searching for Uncle Giant, who was dressed as a diver and rose from the Mersey at Salthouse Dock to a silent reception by crowds, as if he were returning from the *Titanic*. According to Weston (2012), the event was seen over three days by a reported 600,000 people; it was so popular that a second Liverpool procession in 2014 commemorated the centenary of the start of the First World War, featuring a Grandmother puppet in a flowered dress and slippers, who 'slept' in an outsized bed in St George's Hall, before processing with the Giant Little Girl and Xolo to the docks.

The giants are part steam-punk, their mechanism on display and operated by liveried 'Lilliputlian' footmen. One of the authors visited the giant display in Liverpool; the streets of the city were full of people streaming to see the event, waving as if the giants were celebrities, responding to the way in which the puppets are created and the touches of poignancy in their realistic movement, as they seemed to acknowledge the crowd. Royale de Luxe states that 'They are surprising all generations and allowing them to escape from reality for a short time.' The pieces of street theatre are designed as literal landmark events, the magical reality using the cityscape which acts as backdrop and frame for the progress of the characters. The iconic buildings work against an interplay of outsized figures, adding to the notion of hyperreality, when the eye both can and cannot believe what it is seeing and the familiar is made unreal. While the giants perform internationally, Liverpudlians have embraced the event; the word 'heart' was used by the event's directors and the media to describe the response of local people (Coslett, 2012) and the event rootedness. Coslett went on to detail how the event's, artistic director Jean Luc Courcoult, stated:

> I want people to find again the spirit of their childhood when they see these public images. I want there to be a real ownership of the story with the

population of the city. There is a history and past to Liverpool that gives the city an identity and civic pride. It belongs to them and their family.

The 2018 Royale de Luxe event in Liverpool is planned as a further centenary commemoration of the end of the First World War, again combining hyperreality with community to create a magically real event.

Carnival

Perhaps the most obvious aspect of events and the feature that has the greatest impact on transforming the environment is the carnival atmosphere that most, if not all, events and festivals create. They can offer a rustic backstage authenticity as dramatic transformation ensures places are performed. They can be both tangible, in terms of numbers of people and the form of physical spectacle involved, or intangible, in terms of the ambience created. Franklin (2003, pp. 118–19) uses the term 'carnival' to refer to the various ritual festivities that were celebrated across pre-modern Europe such as the festas in Italy, fiestas in Spain, fêtes in France and festivals/carnivals/fairs/revels in Britain, which involved 'a ritual belonging to' and was 'defined by the restricted congregation of a locality'. The carnival was always identified with a particular town and had a number of common characteristics that included, among others, processions, ritual objects, inversions of roles and practices, and 'a heightened party atmosphere generated by more excessive drinking, feasting, dancing and music, but also theatre and games or sports' (ibid., p. 119).

The 'carnivalesque' has implications for the authenticity of self; the term was described by Bakhtin (1984) as a medieval, Renaissance suspension of rules and social class in favour of decadence, echoing the Bacchanalian festivals of the Greeks. The implication is that the social order can be overturned and a more authentic self revealed; Bakhtin emphasised the symbolic use of the folk mask, which suggests an ambiguous duality of persona. Masks are used in a wide range of international rituals and events, from the Haida Raven Dance to hop festivals. When the mask is on, the personality is liberated and the true self revealed. The carnivalesque is associated with the grotesque, involving lewd gestures and slang, licentiousness, ritual plays, theatre, games and foolery, as well as objects, grotesque statues and costume, and the 'mysterious presence (turning up on time and melting away afterwards) of migratory professional entertainers, owners of grotesque side show … travelling theatre companies and so on' (Franklin, 2003, p. 119). As discussed in Chapter 5, carnivals suggest manifestations of magical realism, 'the realistic and fantastical narrative, together with the inclusion of different cultural traditions' (Faris, 2004, p. 1).

Many of the characteristics of the 'carnivalesque' can be seen elsewhere, especially at the seaside, but also at many other modern events and festivals. The annual regatta at Whitby in North Yorkshire, for example, is as much about entertainment as boat racing, if not more so. The event includes fairs, live music and firework displays, together with many of the other characteristics mentioned above and attracts substantial crowds with large numbers of tourists adding to the many residents who turn out for such events. Growing up in Whitby as a child, one of the authors experienced the way in which the regatta and associated activities totally overtook the town for a 24-hour period. The fair would come into town and set up overnight, the stalls and rides would be already operating by the following day, and what had been a relatively quiet seaside resort would be totally transformed, with the streets around the harbour filled with thousands of people experiencing the sights, smells and sounds of the carnival atmosphere. The various regatta races seemed to be almost a sideshow and the experience was further augmented by an evening firework display, with the crowds and heady atmosphere continuing late into the night. By the following morning, however, the fair had packed up and departed and the town reverted to its 'original' state as the mask was removed.

Communitas and heterochrony

Much of the above discussion has referred to the more physical manifestations of the staging and aspects like the carnivalesque, but there are also the less tangible aspects of atmosphere and ambience. People refer to experiencing feelings of enjoyment (Getz, 1997; Ehrenreich, 2007), excitement and 'a sense that something important is happening' (Chalip, 2006, p. 10), which is aligned with experiential authenticity. There is also a feeling that normal life has been suspended and a heightened sense of community spirit. These ideas have been articulated through the concepts of 'ritual', 'liminality' and 'communitas' (Turner and Turner, 1978). The former has already been encountered in Chapter 6 and is more often associated with tourists, as being away from their normal place of residence makes it easier for them to suspend 'normality', but the carnival atmosphere frequently associated with events can engender similar conditions and encourage playful, non-serious behaviour among tourists and residents alike (Bull and Lovell, 2007).

The concept of 'communitas' is a generalised state of friendliness and closeness experienced by those travelling together, but, although originally linked to ritual activities and pilgrimage (Turner and Turner, 1978), it has also been applied to tourists and, in the case of events and festivals, to participants in general. Here it involves the sense of community, albeit on a temporary basis, that is engendered through community cohesion, and participation where the sense of 'unity' refers both to the celebratory 'feel-good factor' and physical mass gathering. Communitas transcends the subject matter of events; it can apply to rock concerts, political rallies, religious events and pilgrimages. The Holy Fire Festival in Jerusalem is an example of an experientially authentic event. It is an orthodox Christian festival, which takes place the day before Pascha at Orthodox Easter and focuses on the miracle holy fire that emanates from the Holy Sepulchre; pilgrims use it to light candles. Participants claim that the fire doesn't burn them.

Like the meeting places described in Chapter 2, events form impromptu stages, performative places of multiplicity and environments of 'eventfulness' (Richards,

2010) where different encounters and flâneurie can take place, a wide variety of people mix and 'out of each unique mixture new processes will arise' (Massey, 2005, p. 62). Many urban spaces were designed for markets, trade and socialising and act as natural arenas for street entertainers and spontaneous on-street performances. Lovell (2013) observed how historic city squares and streets can be transformed through performances into impromptu, subversive theatres, adding multi-sensory qualities to the gaze and rhythms to place. Eventful spaces therefore unfold in unexpected ways, adding elements of 'quantum authenticity' (ibid.).

Drawing on Turner's work, Howell (2013) describes the Badalisc Festival in Andrista, a high mountain village in northern Lombardy, as a folk festival heterotopia, outside conventions. The festival hinges upon the annual ritualised hunt and capture of the Badalisc, a horned serpent creature. Howell describes the 'other place', as that 'which often recurs throughout time and world cultures, and that force us to step out of our everyday patterns, consciousness, and relationships. These universal places, or "emplacements", can challenge and invert our perceptions of "normalcy"' (Sohn, 2008, cited in Howell, 2013, pp. 46–7) and sense of self, but, like the event, the change may be temporary. Berger (2004, p. 54) suggests that 'tourism is a modern attempt… to introduce into the lives of contemporary people the sense of liberation and joy that took place during carnival time'.

A further aspect of Turner's work is ritual. Rituals inscribe and reinscribe events onto places, using emotional energy and shared mood. It is not simply the distortion of space and place that festivals can affect, but also the sense of time, as different eras are manipulated and collective and personal memories reinscribed. Howells (2013, p. 47) draws on the work of Witmore (2006, 2007) to explore the idea of how events can create 'polychronic' or 'multi-temporal' times and places. Polychronism is used to describe archaeological sites, where multiple time periods 'intersect and overlap'. Howells describes how, at a folk festival in the Italian mountains, she was drawn into a sense of mixed time by the sound of a hunting horn, which helped her 'jump into non-ordinary time, into the heterotopia', or 'festival time' when she realised that this was a ritual which has been re-enacted for hundreds of years.

Liminality, communitas and ritual can therefore engender 'personal or intersubjective feelings' (Wang, 2000, p. 49) that are not normally encountered in everyday life, what might also be described as an 'existential state of being'. As Richards (2007, p. 35) argues, citing Wang, experiences of events relate to the 'constructions of authenticity by tourism producers and the tourists themselves (constructive authenticity), but are also mediated by the existential state in which they consume cultural objects (existential authenticity)'. Richards (2015, p. 556) later cites Durkheim's (1915) idea of 'collective effervescence' to describe the mood when people are sharing rituals. Thus, while the designed nature of many events can sometimes seem inauthentic, they nevertheless have the capacity to provide existential experiences that result in real 'other places'. As Cudny (2016, p. 98) describes, 'festivals transform everyday spaces into spaces of festivalisation, which are full of new possibilities, opportunities and activities'.

Constructed identities

As outlined in various other chapters in this text, the interaction between authenticity and place can be organic or induced and theming creates specific cues that are symbolic, such as costume, lighting, colour schemes and displays. Theming makes events meaningful; Bowdin *et al.* (2011, p. 492) state that theming 'is the major artistic and creative decision'. Events are emergent and continually reinvented, their content and design re-scripted for new audiences – for example, the reinvention of classical ballets on more contemporary themes by Matthew Bourne. The importance of a place, especially for tourism, is not simply due to its physical characteristics, but also the way it is perceived. This equates to the conceptual area of second space (Lefebvre, 1991; Soja, 1996). Sometimes, as in the context of the countryside or the seaside resort, constructions of place are strongly influenced by cultural attitudes and traditions and certain festivals and events are also continuations of long-established cultural traditions that will create perceptions of a place. Sometimes these effects have been unintentional, such as the effect of the music festival at Glastonbury. When Glastonbury was first staged in 1970 it was part of a broader movement of outdoor music festivals that began in the late 1960s, the best known being the Monterey Pop Festival in 1967, Woodstock and Altamont in 1969 and the 1968–70 Isle of Wight festivals. Most were one-off events never to be repeated; certainly, no one would have imagined that the Glastonbury festival, which in any case was a relatively small event, would continue to be staged during most years over the next three decades or that it would become internationally famous and one of the 'important socio-cultural and historic landmarks of the 21st century' (Stone, 2009, p. 205). The modern image of Glastonbury, as far as most people around the world are concerned, is now clearly associated with the music festival rather than Glastonbury's other tourist attractions such as Glastonbury Abbey and Glastonbury Tor and their associations with Arthurian legend and the 'holy grail'.

While the effect of some events on image has thus been serendipitous, in other places, as mentioned in the previous section, events have been intentionally promoted because of the recognition that they assist in the development of image (Hall, 1992) and symbolic capital. For example, the Barcelona Olympics, the Madrid European Year of Culture and the Seville Expo that took place in Spain in 1992 constituted a 'mega event trilogy' that provided a way of changing the image of the cities and of Spanish tourism, which had traditionally been associated only with 'sun and sea' holidays (Robertson and Guerrier, 1998). Events and festivals are clearly a key part of the transformation of urban space 'adding new life to city streets and giving citizens renewed pride in their city' (Richards and Palmer, 2010, p. 33). As Richards and Palmer (ibid., p. 31) point out, in addition to the tendency for cities to model their spaces around events, 'many cities around the world have taken this a step further' and now identify themselves as 'eventful cities' or 'festival cities'. Unique events help cities express their 'personalities' (Essex and Chalkley, 1998), they create a favourable impression of the city as a destination worth visiting and 'they provide city marketers with a plethora of suitable vibrant, colourful and multicultural images that are easily packaged for global circulation' (Foley *et al.*, 2009, p. 55).

> Place identity is no longer read off from a city's industrial heritage (e.g. Sheffield and steel, Glasgow and shipbuilding, Detroit and motor vehicles). Rather, place identity is now frequently associated (in the minds of potential visitors, investors and residents, at least) with what it sells, what it creates and what it represents in terms of aesthetic attributes.
>
> Foley *et al.*, 2012, p. 77

Events such as the City of Culture competition – for example, Hull 2017 – do both; they celebrate, reflect and reinforce their industrial heritage, including links to fishing and anti-slavery, which galvanises their sense of place. As Foley *et al.* (ibid., emphasis original) emphasise, 'the cityscape and its "place" attributes is now *the* symbolic commodity being sold to lucrative markets of tourists, investors and workers' as 'events and festivals act to animate the cityscape and generate a plethora of alluring mediated images for touristic consumption' (ibid., p. 87). Emergent authenticity indicates that events change over time and they can become more conventional; Ferdinand and Shaw (2012, p. 8) cite Gay Pride in San Francisco, which at first was an act of resistance, but now has become traditional.

One of the historical causes of hyperfestivity was the perception of Glasgow's image-changing year as European City of Culture (ECOC) in 1990. The positive reading of this event became what was tantamount to a default position, encouraging other cities to bid for mega event competitions in what Hall (2012, p. 128) links to 'grand statements of belief in the discourse of place competitiveness'. The ECOC designation involves a year-long series of events that allow for a much greater impact on the city and image. There has been considerable competition to win the nomination and, although the motives have varied from city to city, most cities have had 'multiple objectives, most often referring to the need to raise the international profile of the city and its region, to run a programme of cultural activities and arts events, to attract visitors and to enhance pride and self-confidence' (Palmer-RAE Associates, 2004, p. 14). From the outset, cities have used the designation to develop their image – for example, Florence, in 1986, explicitly stated in its aims that it intended to use the opportunity 'to reinforce its image'. However, it was Glasgow in 1990 that established the broader, multiple objective approach as the city embarked on using the ECOC as part of a major urban redevelopment programme with explicit reference to both economic and image development aims in addition to strong cultural and social objectives (Richards and Palmer, 2010). According to the Palmer Report,

> a sustained significant long-term impact [of ECOC designation] has been the dramatic transformation of Glasgow's image, from being perceived as a violent, post-industrial city into being celebrated as a creative cultural and leisure centre and one of the most vibrant cities in the UK. Glasgow did not

> have such an image prior to 1990, and the cultural year is widely acknowledged as having a transformative power for the city's image, which has continued undiminished for 14 years.
>
> Palmer-RAE Associates, 2004, p. 167

Glasgow City Council has continued to see the hosting and staging of events as an important part of its strategy, illustrated by their Winter Festival, which provides a range of events starting with Guy Fawkes and ending with 'Glasgow's Hogmanay', and the Celtic Connections' folk music festival. Foley and McPherson (2007, p. 154) offer the observation that 'with the establishment of the EventScotland agency, it seems that selling Scotland and its culture as a commodity is becoming more important for cities, particularly in terms of their cultural and regeneration strategies'.

Despite its success on one level, there has been some criticism that the Winter Festival was a further example of a policy designed to portray a reinvented image of Glasgow that excluded certain types of Scottish and Glaswegian culture. This is a similar criticism to that voiced after the ECOC collection of events by Garcia (2005, p. 845), where a very positive press view of the city's image transformation and economic legacies in general and the growing levels of tourist visits were countered by academic publications 'dismissive of the rhetoric surrounding 1990' (Booth, 1996; Boyle and Hughes, 1994; Mooney and Danson, 1997; Spring, 1990). He also quotes Harvey (1989b, p. 21) who argues that prestige arts-led and related large event regeneration projects function as a 'carnival mask that diverts and entertains leaving the social problems that lie behind the mask uncared for'. Johansson and Kociatkiewicz (2011, p. 12), in their study of festivals in Stockholm and Warsaw, repeatedly refer to urban festivals resulting in aspects of the city being 'sanitised' and suggest that 'as well as presenting a sanitised version of the city's cultural landscape, the festival also serves to hide from view the less palatable elements of the urban experience'.

Conclusion

This chapter centres on how events function as masks, temporarily concealing and revealing the authenticity of places behind different personas and aura. Events and festivals can act as placemakers, transforming places into new, exciting and vibrant staged and lived environments, creating their own sense of occasion and reality. Tourists use events to experientially authenticate places; even staged authenticity can generate real feelings of communitas, backstage insights and inclusivity. On a macro level, in places such as cities, their transformative impact can be substantial and permanent, leveraging new investment, infrastructure and social capital; gentrifying both the fabric and image. However, not every community shares the benefits, legacy and transformation. Events often draw on the local traditions of places and, on a micro level, where a 'slow' planning process is applied, their veracity is strengthened and more strongly located through local distinctiveness, products and people. While many events are design-led, re-enactment events emphasise the granular authenticity, rather than the staging. At the extreme end of this scale, some peripheral communities exclude tourists from the area, in an attempt to de-commodify their events.

New technologies create the potential for increasingly immersive, hyperreal and fantastical forms of expression that exaggerate aspects of places. Boorstin and Debord's concepts bear revisiting in the age of social media, when events are designed to be consumed both physically and virtually and to attract as much attention as possible. Hyperfestivalisation increases the threat of standardisation and hyper-replication of places, but it could be that an over-supply of events is necessary to feed the insatiable needs of the virtual and physical experience economy.

The chapter has suggested that there are multiple realities evident in events that blur ideas of fake and real; festivals have been described as heterotopias that can simultaneously juxtapose an invented and a real space. Festivals also display attributes of heterochrony, evoking their own timescales and mixing past and present in heritage re-enactments.

Although the resulting event-places are transient and may seem inauthentic and standardised, suggesting and designing meaning in a choreographed manner, participants use events to experientially authenticate places. Many events create a carnival atmosphere, with its associated ritual and associated elements of liminality and communitas (Turner and Turner, 1978) inducing opportunities for spontaneous, unique, personal experiences, experiential authenticity and magical realism. In the case of the carnival, the mask itself seems to reveal the true nature of people and place.

11 Variations on a theme

The imagineering of fast authenticity from theme parks and museums to shopping malls

Introduction

The geography of inauthenticity could be argued to be particularly evident in themed locations. Getz and Page (2010, p. 237) define 'themes' as unifying ideas and 'themed environments' (Shaw and Williams, 2002, 2004; Urry, 2002) are locations which are connected by a design concept. The chapter explores a range of places physically constructed to create artificial (or falsified) locations which are based on a theme or a fantasy. They are designed by 'imagineers' (Fjellman, 1992) who specialise in creating a phenomic, immersive environment which blurs fantasy and reality. At their most vivid they are operatic and super-strength; the exemplars are the various Disney theme parks and the 'global pleasure zone' of Las Vegas (Gottdiener, 1997 [2001]) described by Hannigan (1998) as a 'fantasy city'; but they also include themed shopping malls and regional divisions of countries based on associations with historic and literary figures and television/film locations (e.g. Robin Hood country, Thomas Hardy's Wessex, Poldark's Cornwall). The chapter also observes how some museums have been compared to theme parks, evident in their techniques of merchandising, interactive technology, digitisation and the use of social media which ensures that the objects which they contain are reproduced.

On one level, theme parks correspond to several of Fjellman's (1992) typologies of Walt Disney World, fake fake and real fake. Lovell (2013) defined a range of real-fake places – for example, shopping malls and gentrified areas – where the emphasis is on the staging rather than the authenticity. Furthermore, theme parks resemble Benjamin's (1955) phantasmagorias, particularly the modernistic elements which Jamal and Kim (2005, p. 66) refer to as 'artifice, calculated illusion, metronomic spectacle and automated simulation'. Baudrillard (1981, p. 166) states Disneyland is the perfect simulacra; 'it is first of all a play of illusions and phantasm'. Theme parks are third-order simulacra; they are exaggerated copies of fantasy worlds with their own fast originality.

At first sight, these are not original places designed to be authenticated, but, as the discussion of the Wizarding World of Harry Potter theme park in this chapter demonstrates, their design is not post-authentic or post-truth, but complies strictly with the details and rules of the fantasy universe, as set out in the books expected by their 'muggle' visitors; providing an example of 'imagineered authenticity'. Like some events, many theme parks reflect Debord's (1983) *Society of the Spectacle*, where the commercial aspects are concealed behind signifiers and markers. Despite the obvious construction of some themed places as fake – rescaled representations of original places, simulacra and three-dimensional recreations of filmic places – and their acceptance as such, they may nevertheless be valued for their inauthenticity and have their own essence and rules. In most theme parks commerciality is on display, not leaving much to the imagination and ensuring that you exit through the gift shop. Lovell (2013) explored how visitors to historic cities identified as 'fake' signs of commodification created opportunistically for tourists, suggesting they were 'hotly inauthenticated' (adapted from Cohen and Cohen, 2012). In the context of theme parks, the search for inauthenticity is key; people expect and actively search for those artificial, commodified souvenirs and fast food as signifiers of sense of place; thus theme park merchandise could be said to be authenticated by tourists.

Overall, while they may seek to retrace the steps of previous visits and re-authenticate memories, tourists are less concerned with authenticating originality, but instead preoccupied with pursuing a 'playful search for enjoyment' (Cohen, 1995). The search for fun within these themed attractions may also afford opportunities for magical realism, where, for brief moments, fantasy assumes an everyday form. While people can still distinguish between reality and fantasy, the popularity of virtual reality demonstrates that sometimes they choose not to do so. Theme parks use atmospherics to evoke an immersive, heightened state of hedonic consumption for thrills which comply with experiential authenticity and blur the real and fantasy into the sensational. Some of these themed environments, especially theme parks and malls, may now be so important in people's lives that they have moved, as Lovell (2013) describes, from 'fake to real' becoming an established (and thus authentic) element of contemporary culture. Authenticity may not be actively sought in many themed places, but it may certainly be encountered.

Theme park 'fast authenticity'

Although many different types of theme park exist throughout the world, it is the Disney parks that most epitomise the concept. The first Disney theme park was opened in 1955 just outside Los Angeles in California and, according to Shaw and Williams (2004, pp. 244–5), was fundamentally different from other leisure spaces in a number of ways. First, visitors do not pay for individual rides but for admittance to the park itself and thus for the experience of the total built environment and associated fantasy. Second, visitors 'enter a carefully constructed, heavily regulated and well-choreographed space' in which they are immersed in a fantasy (but safe) environment that 'provides entertainment and excitement, with reassuringly clean and attractive surroundings' (Smith, 1980, p. 46, cited in Shaw and Williams, 2004). Third, Disney-type theme parks provide a strong contrast to the spaces of everyday life, especially the proximity of Disneyland to Los Angeles, and Shaw and Williams (2004, p. 245) list several of these based on Gottdiener's (1997 [2001]) observations of American life. These include:

- Spaces that can be enjoyed by pedestrians in contrast to the dominance of the car in suburban America
- Safe, regulated environments free from crime, unlike the wider experience of urban America
- A festival-type environment, with both active and passive forms of entertainment which contrasts with the dominance of passive entertainment in many people's suburban lives
- A seemingly liberating experience, with its illusions of escapism from the demands of everyday life.

The appeal of such contrasts accords with MacCannell's (1973, 1976) view that tourists are always motivated to travel in search of the authentic, driven by a need to escape the humdrum reality of their everyday lives and experience something different. MacCannell believed the search to be illusory, with the result that visitors were only able to experience a fake reality, what he referred to as 'staged authenticity'. According to Pikkemaat and Schuckert (2007, p. 199) 'it is obvious that theme parks are able to fulfil tourists' needs for authenticity' in that they 'seem to be able to deliver both personal fulfilment and escape to other places'. Of course, as outlined in the introduction, there is no simple division between the authentic and inauthentic. As Wang (1999, p. 357) points out, 'in WDW [Walt Disney World]there is no absolute boundary between the real and the fake'. The real may turn into the fake and vice versa and the 'lines between the real and the fake are systematically blurred' (Fjellman, 1992, p. 255). As Lovell (2013) explored in the context of historic cities, tourist states of authenticity are permeable and fluid. Wang (1999) also refers to work by Cohen (1995), who argues that postmodern tourists have become less concerned with the original, pursuing, instead, a 'playful search for enjoyment' or an 'aesthetic enjoyment of surfaces'. Furthermore, Cohen (2007, p. 78) also cites Disneyland as an example of 'emergent authenticity'. As he points out, what is generally regarded as 'the initially quintessential, overtly "framed" tourist attraction, has over time been increasingly recognised as part of contemporary American culture, and as such, is authentic'.

In addition to the original Disneyland, replica parks have since been opened in Florida (Disney World), Paris, Tokyo, Hong Kong and Shanghai and all contain similar themed spaces (or lands) and thrill-seeking rides. The California Disneyland website, for example, describes:

> a magical kingdom where you can sail with pirates, explore exotic jungles, meet fairy-tale princesses, dive under the ocean and rocket through the stars – all in the same day! Disneyland Park is a beloved California destination

> where generations of families have made their Disney dreams come true.... Walt Disney's original theme park is divided into 8 extravagantly themed lands – Main Street, USA, Tomorrowland, Fantasyland, Mickeys Toontown, Frontierland, Critter Country, New Orleans Square and Adventureland.

These nostalgically themed areas represent features developed in films and television programmes made by the Disneyland Corporation. 'Main Street', for example, is the first 'themed land' that visitors experience on entering the park and represents a typical American town at the beginning of the twentieth century. It incorporates turn-of-the-century architecture in small-town Middle America and is described as the 'crossroads of an era' when gas light was giving way to electric light and both horse-drawn cars and motor cars could be seen on the street. Frontierland, another of the original themed spaces, is a re-creation of the Old West where, among other things, visitors can experience the thrills of a runaway train ride, enjoy a leisurely riverboat cruise and watch an Old West stage show. According to Shaw and Williams (2004, p. 245) it was based on representations of early settlements, as highlighted by old Disney film classics such as *Davey Crockett* – about the American frontiersman.

Gottdiener (1982) describes Disneyland as a 'utopian urban space' which 'presents the illusion of cornucopia' (p. 152) and 'the myth of small-town America' (p. 158). Citing Lefebvre (1974, p. 423), he suggests it possesses 'the "illuminating potentiality" of a space occupied by the symbolic and the imaginary in which something fantastic can and usually is always happening' (p. 154). He goes on to suggest that each of the various themed areas corresponds to 'compartmentalised aspects of the world of a young boy growing up in a Midwestern town' (p. 157), probably based on Disney's own childhood and his knowledge of fables and nursery rhymes and his small-town experiences.

Despite its provision of excitement and perceived adventure, Disney theme parks also involve a subtle, controlled environment. As Eco (1986, p. 4) suggests, this involves 'access to each attraction [being] regulated by a maze of metal railings which discourages any individual initiative' and involves 'officials of the dream, properly dressed in the uniforms suited to each specific attraction' regulating every move relating to queuing, sitting down, standing up and moving on. As Eco (ibid.) stresses, 'if the visitor pays this price, he can have not only "the real thing" but the abundance of the reconstructed truth'. Shearing and Stenning (2002, p. 431, cited in Hannam and Knox, 2010, p. 24) also describe this 'instrumental discipline', suggesting that control within Disney World is 'embedded, preventative, subtle, cooperative and apparently non-coercive and consensual'.

As the Disneyland theme park became transplanted into new destinations, not only have the themed environments within these parks changed and become more complex and elaborate (e.g. the Epcot centre in Disney World, Florida, with its laser-light shows and world showcase, represented by mini-themed areas of 11 countries), but also a variety of different theme parks have emerged. According to one observer:

> Disney World has expanded, in typically orderly fashion, one module of imaginary world at a time, becoming not a city that is a theme park, but a theme park that has become a city. Disney World has even developed its own suburbs of fantasy, that are filling central Florida with theme park sprawl, as miniature and not-so-miniature attractions, featuring Medieval knights, re-created Chinese buildings, and an animatronic King Kong, spring up around its outskirts.
>
> transparencynow.com, 2016

The Disney parks are some of the most popular attractions in the world, with Disney World's Magic Kingdom receiving over 19 million visitors and Disneyland Paris over 10 million visitors in 2013 (TEA/AECOM, 2015).

It is worth noting that the artist Banksy's nightmarish Dismaland theme park in north Somerset in the summer of 2015 reframed the Disney phenomenon, creating a dystopian visitor attraction, where smiling was forbidden. The Disney-style castle was unfinished, Cinderella's carriage was overturned in an accident with pursuing paparazzi, the grim reaper rode on a bumper car. It could be argued that this was a more realistic version of a theme park and was designed for those 'looking for an alternative to the sugar-coated tedium of the average family day out' (Garcia, 2005).

There are plans for several new, European mega theme parks which would add a scale of the theming and heritage already on offer. In the UK, for example, London Paramount is intended as a mega theme park in Ebbsfleet Garden City, reflecting the broad spectrum of Paramount films and featuring UK productions such as the Aardman, Harry Potter and James Bond franchises. Many of these films reflect a certain type of Britishness and, if the theme park receives planning position, its effect on perceptions of the UK identity will be global. If approved, 'the Entertainment Resort will contain many themed zones, with exciting rides and attractions for families, children and the more adventurous thrill-seeking visitor' and themed restaurants, such as a Godfather eatery (featuring, one hopes, cannoli on the menu) (London Resort Company Holdings, n.d.). The company goes on to state that it expects to receive 40,000 visitors per day.

In a similar vein, but with a historical emphasis, Norway plans to open the Aasgard theme park, located on Haugesund and capitalising on new interest in Norse mythology in many of the blockbuster films such as *Thor* (2011). As Russell states (2013)

> 'This area is known as the homeland of the Viking kings,' says Odd Erik Salvesen, project co-ordinator at Vikingland Utvikling AS. 'We've got a load of history here, including the remains of King Harald Fairhair – the man who reunited Norway. So we wanted to see if there was a way to capitalise on this and attract more people to the region.' Early artists' impressions of Asgard-Viking Adventure Park show mythical swirling towers, turquoise-lit

> waterfalls and a plethora of pastel colours. So are we about to see the Disneyfication of the Vikings? 'We admire the Disney model,' admits Salvesen, 'but there will be less plastic and more natural materials in our Viking land. We're aiming to create a Viking-atmosphere by locating the park in an environment that's pretty much as it was in the Viking-age.' Only with more ice-cream vendors and light shows, presumably.

The Viking Land plan is to draw on the images of Norway, interweaving authenticities – for example, those from real history, Norse myth, films of Norse myths, living history and themed atmospherics.

At the core of some theme parks is the rubric of their own storyworlds. For some theme parks their reality involves conforming to fictional universes, which is achieved through the persuasive imagineering of mechanistic authenticity. Hogwarts and Hogsmeade are fictional Harry Potter destinations, identified as Scottish locations in the film versions, with its lochside settings and the Fort William to Mallaig steam train. The Wizarding World of Harry Potter reproduces Hogwarts at Universal Studios Florida and Hollywood and Universal Studios Japan, each reflecting their local distinctiveness with the fictional place. As Universal Studios (n.d.) state:

> Hogsmeade has been faithfully recreated in three locations across the world. Each land has its own characteristics… and attributes that are tailored to their locations, a challenging task when you consider the importance of remaining true to the films. Alter too much, and you'll begin to distort the authenticity of J.K. Rowling's fanciful world of Harry Potter. Alter too little, and you may get a land that doesn't appear to take advantage (or even acknowledge) its surroundings.

To differentiate locations they use local attributes, such as Californian sunshine to create shadows. As Universal Studios go on to state, 'And, of course, come sunset, nothing can beat a photograph of Hogsmeade with a California sunset as your backdrop' (ibid.). The theme park creators are clear that there is a distinction between worlds, which are kept parallel, but sometimes they must converge.

> Whenever possible, Universal sought to immerse guests into the Wizarding World without tearing them out of their element…. you can sometimes see the 'real world' from inside Wizarding World Hollywood, but 'where we can, we will bring you into Harry Potter's world as best we can.'
>
> (Ibid.)

The story universes also conform to heritage standards and have a placebo effect. Film theme parks, such as the Warner Bros. Studio Tour 'The Making of Harry Potter' in London, offer an experience of collections which reify memorabilia in a similar way to historical museums and, as the author can attest from a visit, the experience of reverence is similar. The film artefacts are designed, presented and consumed like museum pieces, and are examples of 'fast authentication'. As Alan Gilmore (2016), Supervising Art Director described on the 2016 opening of the Wizarding World of Harry Potter in Los Angeles,

> We do create a world that is the same as the books and the films. There is a real logic behind what we do here. Every detail actually comes from a real place, like this chair is gothic, so people even though they know it's new, they accept it as being here for a long, long time.

The celebrity dust attached to a filmed item is equivalent to a historical artefact that has witnessed events such as battles. Both fame and the distance of time confer a sense of aura, establishing a distance that can never quite be dissolved.

These parks are a global phenomenon; in addition to the Disney parks in Hong Kong and Shanghai, China also has various other theme parks. In direct opposition to Disney, the Chinese conglomerate, Dalian Wanda Group, established two new theme parks at Nanchang and at Hefei in 2016 and has plans to open 15 similar mega-resorts by 2020. The design of the Hefei park gives centre stage to elements of the local Hui culture in keeping with the company's philosophy of wanting to 'move away from Western imports and to establish a global brand based on Chinese culture' (BBC News, 2016).

The emphasis on Chinese culture can also be seen in other Chinese theme parks. The Millennium City Park in Kaifeng, for example, is described as a historical and cultural theme park which recreates *Riverside Scene*, a world-famous painting by Zhang Zedun, a well-known painter of the Northern Song dynasty. According to the Kaifeng Travel Guide,

> when you step into the park, you will feel like you are travelling back to the flourishing Song Dynasty. Large and grand ancient Song style buildings and staff wearing ancient Chinese clothes make the park full of ancient charm and elegance.... The whole park is arranged exactly like the original layout of Riverside Scene with the Jiu Lou ['wine buildings', like restaurants today], teahouses, pawnshops, and the manufacture of Kaifeng embroidery, porcelain, new year pictures and so on.

In addition to its physical environment, the park also provides over 20 performances throughout the day that allow visitors to experience the life and customs of the Song dynasty. Similarly, the Songcheng theme park in Hangzchou also features cultural characteristics of the Song dynasty based on the same painting.

It is important to complete this discussion by mentioning the Legoland theme parks, which were created by the Danish company, Lego. Lego (see Figure 11.1) has specialised in creating what could be termed a 'lego reality', incorporating versions of places and of films; research shows that tourists appreciate the fun the park provides (Bakir and Baxter, 2011). In the Legoland theme parks, places are

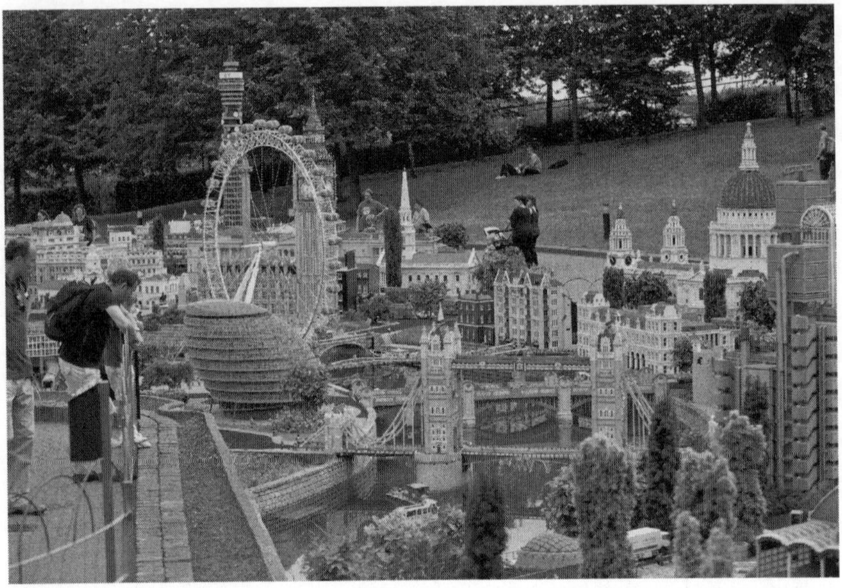

Figure 11.1 Miniland Legoland
Source: Legoland

reproduced in miniature and worlds are compressed still further: from Venice in the Legoland in Germany to Las Vegas in the Legoland in San Diego.

Las Vegas

While Disneyland is often cited as the classic example of Eco's (1986) 'hyper-reality', Las Vegas is equally deserving of such a claim. Venturi *et al.* (1972, cited in Selby, 2004, p. 106) demonstrated how Las Vegas created the ultimate hyper-reality, with its landscape mixes of Egyptian temples, reconstructions of the Wild West and Victorian England, while Gottdiener (1997 [2001], p. 100), commenting on its more recent structures, describes it as both a 'global pleasure zone' and a 'fully themed environment'. Shaw and Williams (2004, p. 248) explain how this transformation during the 1970s from a Western-style gambling resort in the middle of the Nevada desert was based on 'the creation of a well-planned themed environment where fantasy and pleasure are unifying motifs in the landscape'. They suggest that the fantasy space is 'multifunctional offering visitors gambling, sex, food and nightlife' and that 'the gambling economy… is embedded within structures that are a complex series of themed resorts, which both compete with and complement one another'. Hannam and Knox (2010, p. 53) describe how Las Vegas contains numerous replicas of other world tourism sites and quote Smith and Bugni (2002), who suggest that:

> Everything that is supposedly enjoyed and experienced can be encountered on the grandest of scales in Las Vegas, including the New York Skyline, the great monuments of Paris, instantaneous escapes to Rio, the exotic lure of Asia at Mandalay Bay, the pyramids of ancient Egypt, the Caesars, queens, and gladiators of ancient Rome, sinking pirate ships, remembrances of early Hollywood, and duelling knights of the Roundtable. The belief is that anything that can be imagined can also be designed and accomplished in Las Vegas.

The 'fantasy space', which, in 2012, welcomed almost 40 million visitors (Las Vegas Convention and Visitors Authority, 2016), is concentrated along a 7km stretch of Las Vegas Boulevard, often referred to as 'the Strip', and contains numerous casinos and resort hotels (some of the largest in the world), shopping malls, restaurants and bars, nightclubs, theatres and a variety of other tourist facilities, with many of these attractions being integrated within a themed environment. According to Raento and Flusty (2006, p. 99) 'the Strip compresses world history and geography in one place for our convenient entertainment'; they suggest that Las Vegas facilitates Zelinsky's (2001) claim that people are capable of transforming themselves 'anywhere and anywhen mentally', instead of being 'tied to the here and now' and that there is also a pronounced craving for 'imagining and creating alternative existences'. Like Disney, Las Vegas provides experiences in a clean and safe environment and the underlying mechanisms that provide the entertainment product involve 'the manipulation of scales, time, space, and sense of place', as well as 'the manipulation of sense of distance and spatial perception, boundaries, freedom and control' (Raento and Flusty, 2006, p. 121).

One of the most famous examples of such developments is Caesars Palace. Its luxury hotel and casino complex was established in 1966 and was designed to provide guests with a sense of life in the Roman Empire, with its many statues, columns and iconography typical of Hollywood films of the Roman period. A more recent example, opened in 1999, is the Venetian Resort Hotel Casino, which comprises a luxury hotel and casino resort. The complex involves architectural replicas of various Venetian landmarks, including the Palazzo Ducale, Piazza San Marco, Piazza di San Marco, the Lion of Venice Column and the Column of St Theodore, St Mark's Campanile, and the Rialto Bridge. Attractions also include indoor and outdoor gondola rides, which further add to the themed experience.

The casino resorts of Caesars Palace and the Venetian, along with the Bellagio, are considered by Raento and Flusty (2006) as 'three versions of Italy' in their study of the new Las Vegas. Using these as detailed case studies, they demonstrate how their locations afford convenient journeys in both space and time, with ancient Rome being just across the street from medieval/Renaissance Venice and the Bellagio, located next door to Caesars Palace, providing a rural contrast. As they explain, the Bellagio involves a 'moving walkway to the entrance' that offers:

> a view to the nine-acre lake in front of the property and the picturesque 'Tuscan village' on its shores. At the walkway's end the Bellagio's wall paintings of bucolic landscapes suggest sophistication, unspoilt rustic and romantic peacefulness. This is where the city returns to the countryside – where the elegant aristocrats withdraw from the heat, dust, and noise of the city.
>
> Raento and Flusty, 2006, p. 108

The theming experience at these 'resorts' is also enhanced by other devices such as the suspension of time and space. For example, in certain spaces ceilings are painted/decorated to produce artificial skies which can give the illusion of diurnal change. Raento and Flusty (ibid., p. 107) describe their experiences of such artificial skies that 'can control perspective and perception by making the clouds appear as if they were moving when one walks underneath' and where controls can produce a sunset at 8.45pm, followed by a night lasting 20 minutes, followed by a 40-minute day and then another sunset. A further aspect of theming involves the various artefacts and souvenirs that are sold in the many boutiques. In the Venetian, for example, the decorative items and jewellery pieces evoke the traditions of Venetian manufacturing and retailing and 'crystallize the strong association between an individual artefact and the place they represent by bringing together Venice's romanticised reputation, landscape and lifestyle' (ibid., p. 116). Theming is also further enhanced by certain key workers wearing period costumes – for example, at Caesars Palace staff dressed as Roman aristocrats and soldiers entertain the visitors in the casino.

While the theming is intended to produce an authentic replica of the place, the reality is a hyperreal experience with scale compressed to afford convenience of travel and the less attractive elements, the grit, removed in acts of 'structured absence'. For example, at the Venetian there are none of the unpleasant smells, pigeon droppings and rubbish that would be found in medieval Venice, and the routeways are designed to enable passage by everyone, including those with disabilities. The themed environments are designed to accord with visitors' idealised expectations and with what they prefer to exclude from their tourist spaces.

While some of these authentic, unwanted, realities are completely absent, there are other aspects which are conveniently hidden from the tourist gaze. Visitors are almost entirely confined to the Strip, which operates as a type of tourist bubble, sheltering them from the rest of the urban area that houses the hundreds of thousands of workers who service their needs and who maintain the façade and attractions (Greater Las Vegas has over 2 million inhabitants). This aspect is particularly highlighted in Littlejohn's (1999) text *The Real Las Vegas: Life Beyond the Strip*, which looks at the lives of the workers and their families who inhabit the city, including teenagers (whose suicide and dropout rates frighten parents), senior citizens (many of whom spend their days playing bingo and the slots), Mexican immigrants (who build the new houses and clean the hotels), homeless people and angry blacks, as well as local police, active Christians, city official and prostitutes.

Similarly, Rothman and Davis's (2002) edited collection, *The Grit beneath the Glitter: Tales from the Real Las Vegas*, also depicts 'the other side of Las Vegas' with the various contributors uncovering the contradictions between the illusion and the reality of the city, the seam between fantasy and the life it masks. Many of those who work on the Strip are low skilled and low paid and their lifestyle provides a stark contrast to that of the affluent tourists whom they serve.

A further issue to which most tourists remain oblivious is the problem of water supply. The lake outside the Bellagio and the fountains (which repeatedly shoot 500 feet in the air), together with the Venetian canals, the Roman themed pools and the thousands of sprinklers that keep the golf courses and lawns green, all combine to give the illusion of an endless supply of water. And yet the supply of water for Sin City is extremely precarious. It relies on water from the Colorado river feeding into Lake Mead reservoir at the Hoover Dam. However, due to the continual building in the city and a 14-year drought, 4 trillion gallons of water have been drained from Lake Mead in the last decade, resulting in it being only half full (*Telegraph*, 2014). Las Vegas receives only 4 inches of rain in a good year and yet it uses 219 gallons of water per person per day; while a variety of schemes and practices are proposed to solve the water supply problem, it remains a potential crisis with even the future of the city and its economy at stake. Clearly, the appearance of a fully functioning and seemingly temperate environment in the middle of a desert is one more of the many illusions that characterise Las Vegas, the city which, paradoxically, contains the 'Mirage' chain of hotels (Gottdiener, 2001, p. 108).

The design of Las Vegas is so fake it becomes real as a reflexive metaphor for consumption, a joke which we all share. Yet it is an example of emergent authenticity; it was described as 'a fake becoming a real place' by Huxtable (1997, p. 75), suggesting a process of 'fake to real', as an invented tradition or place becomes an accepted part of the fabric.

Themed shopping malls

Themed spaces and theme park attractions can also be found at various large shopping centres. In recent decades shopping has become a major leisure and tourism activity (Jansen-Verbeke, 1990, 1991; Goss, 1993; Falk and Campbell, 1997; Timothy, 2005) and there has developed:

> a blurring of the distinction between tourism precincts and the large suburban shopping centres which increasingly cater for visitors as well as residents with a combination of retail, entertainment, souvenir and gallery establishments and nearby accommodation for holiday and business travellers.
>
> Kelly, 2008, p. 121

In attempts to further enhance the attractiveness of the overall leisure and shopping experience, some centres have developed themed spaces. In the same way that Las Vegas has created themed attractions to attract more consumers and gamblers to its casino complexes and Disney uses themed retail outlets to enhance its merchandising, so too have shopping malls established falsified, geographical places to encourage consumption. According to Lehtonen and Mäenpää (1997, cited in Shaw and Williams, 2004, p. 126) 'a trip to the modern shopping mall is, for many consumers like visiting somewhere else, where the real world is challenged by the possible world', although as Butler (1991, p. 291) observes in relation to West Edmonton Mall in Canada, the emphasis is on 'selling an ambience of foreignness rather than specific locations'.

West Edmonton Mall is perhaps the epitome of this type of development being not only the first so-called mega-mall when it was opened in 1981, but also because of its large size. Even today it remains the largest mall in North America and is the tenth largest in the world, attracting over 30 million visitors a year (TEA/AECOM, 2015). It contains a variety of major tourist attractions, theme streets, over 800 stores and services, 26 movie theatres, hotels and activity centres, including an indoor lake and outdoor parking for over 20,000 vehicles. The major attractions include Galaxyland (an indoor amusement park – the second largest indoor amusement park in the world); World Waterpark (which includes the world's largest indoor wave pool); Sea Life Caverns/Sea Lions Rock (which contains a salt-water habitat with a sea lion and a replica of the *Santa Maria*); Mayfield Toyota Ice Palace (which is a scaled down version of a National Hockey League ice rink); and Professor WEM's Adventure Golf (an 18-hole miniature golf course). It covers a gross area of about 490,000m^2, employs over 24,000 people and attracts over 30 million visitor per year (*Edmonton Journal*, 2015). Once promoted as the 'Eighth Wonder of the World' (*The Canadian Encyclopedia*, 2015), it has attracted much academic interest as a major phenomenon in tourism (Butler, 1991; Jackson and Johnson 1991; Goss, 1993; Finn and Erdem, 1995; Timothy, 2005). It displays many comparisons with Disney World and, according to Page and Hall (2003, p. 137), it is a reflection of postmodernity and 'represents a landscape of spectacle, fantasy, pastiche and artificial creations of new shopping environments'. The theming is all pervading and, as Shaw and Williams (2002, p. 206) point out, the 'visitor can be in France, England, Asia, or West Coast America – all within Fantasyland (now renamed as Galaxyland)'. They also provide a list of the various themed 'falsifications of place' (based on Butler's [1991] study), which includes representations from Europe (England's crown jewels, Versailles fountains, a Roman room and Caesar's statue), North America (e.g. Las Vegas' Caesars Palace bingo, Bourbon Street, New Orleans, rafting at Grand Canyon), Asia (Ming vases and rickshaws and Japanese pagoda), Arabian and Polynesian themed rooms and representations of the ocean through a dolphin show, aquaria, submarines and the *Santa Maria*.

Another similar and well-documented mega-mall is Mall of America, which was opened in 1992 in Bloomington, Minnesota, USA. Like West Edmonton Mall, it includes various themed spaces and attractions. Goss (1993) provided a detailed description and examination of the development, especially its strong themed 'nature'; an excerpt of this is quoted here:

> An essentialist Minnesota 'sense of place' pervades the Mall, from the main attractions to retail and restaurant themes. Camp Snoopy for example (from the Peanuts cartoon created by Charles M. Schultz, a famous native son of St Paul), is the 'nation's largest indoor theme park'... and the Mall's major attraction. It evokes 'a rustic clearing deep in Minnesota's northwoods'... and convincingly reproduces a stretch of a northwoods stream. A Northwoods feeling is evoked by 30,000 plants and trees – the largest indoor planting in the world – an artificial seventy-foot waterfall, plaster cliffs cast from the originals along the St Croix River, browsing animatronic moose, and bird noises... UnderWater World, the second major attraction, is an aquarium that offers the prospect of 'diving' into Minnesota Dawn, a cool glacial lake with morning mist, beaver dams, and local fish. Even where the regional theme is not directly evoked in the names of stores and restaurants, northern landscapes and ways of life are employed in the mise-en-scène, and retailers sell authentic souvenirs of the Mall's Minnesota experience.
>
> Goss, 1993, p. 50

The mall continues to mount themed exhibitions on a vast and detailed scale, such as a 2017 *Downton Abbey* costume display. Themed malls are also found elsewhere in the world. Cité Europe, near Calais in northern France, offers visitors 'the chance to become an international shopper' with a 'cosmopolitan choice of shops from around Europe in one huge space' (Calais-guide, n.d.). Several mega-malls can be found in the Middle East. The Dubai Mall, for example, is the world's largest mall in terms of area (12.1 million sq ft) and contains themed shopping areas like the Gold Souk, Fashion Island and The Grove, an indoor–outdoor streetscape with a fully retractable roof.

Butler (1991) argues that mega-malls create an image of 'elsewhereness', while Shaw and Williams (2004, p. 126) suggest they 'provide safe, controlled environments for consumers to spend their leisure time, merging acts of consumption into notions of play'. They therefore contain many of the same attributes of Disney theme parks and, like Disney, they have also become as essential part of contemporary culture. Thus, they may also represent a further example of Cohen's (2007) 'emergent authenticity', as well as providing visitors with a variety of potential existential experiences.

Museums

This chapter has so far concentrated primarily on large-scale themed attractions, but there also exist many other smaller examples. While museums may not possess the thrill-seeking rides of the theme parks or the extensive retailing opportunities of the malls, they nevertheless still provide visitors with a themed experience which, given usually a more specific focus, may be equally intense. At first glance, many museums and galleries don't resemble theme parks, but cultural tourists are likely to visit both types of visitor attraction. Some national museums, such as the British Museum and National Gallery are built in imposing classical styles as muscular statements of national power (Tuan, 1974), 'mighty real' (York, 2016) designs reminding the viewer of the seriousness of art enclosed within these temples of culture. National galleries such as the Musée D'Orsay in Paris and Tate Modern in London have repurposed iconic industrial heritage buildings and museums. In contrast the National Museum of African American History and Culture in Washington used a 'starchitect' to create a prototypical modern building. The iconic structures are intended to be arresting and impressive, emphasising the aura of their contents; however, museums were pejoratively compared to theme parks in the tourism literature of the 1990s, due to their increasing emphasis on interactive media (MacDonald and Alsford, 1995; Huxtable, 1997, p. 44; Hewison, 1987; Walsh, 1992; Kirschenblatt-Gimblett, 1998). Smith (2003, p. 24) encapsulates the observation in the following quotation:

> a number of authors have criticised the way in which heritage sites and landscapes have become more and more like theme parks. This is largely due to the post-modern pastiche of reality and fiction, and history and media, which can result in a kind of simulacra, interrupting the flow of time.

The business of museums may seem to be fostering authenticity and original artefacts, but they also specialise in making reproductions that have their own value – for example, the plaster skeleton of Dippy the Diplodocus Dinosaur in the Natural History Museum, which went on tour in the UK in 2017. Reassessing the relationship of museums to theme parks, many modern museums extend their skill to reproduce and recreate, in three different ways: first, their commercial approach; second, their use of social media and immersive digitisation; and, third, their use of merchandise. Their entrepreneurial approach recognises that the objects which they hold are copyrighted and rare; they are using reproductions of merchandise and digital copies to increase their value. Padilla-Meléndez and del Águila-Obra (2013, p. 7) observe that:

> Digital technologies create new possibilities for museums to exploit their cultural assets and create more value, allowing them to deliver and assess customer value in new ways; for example, by digitising their collections and archives or providing new services to existing customers, they are able to expand audience reach.

In this way, museums can break their 'traditional value chain' (ibid.), indulging in initiatives such as Museum Selfie Day, locating visitors within and as part of the artworks, setting up shots making it look as if the statues and paintings are animated. The Rijksmuseum in Amsterdam has placed high-resolution images of its collections online in the RijksStudio. This means that people can use the image to decorate a car or a mobile phone cover, authenticating everyday objects as artwork. The museum is not daunted by the fact that the proliferation of their images leads to their personalisation and to changed meaning, which – even when altered – does not detract from the aura of the images that are still covetable and rare.

Museums employ social media to extend their global reach into virtual reality: As Fletcher and Lee (2012) discuss, museums use social media increasingly effectively, which encourages people to display their personal cultural capital. Apps such as that at the Musée du Louvre enable people to access interpretation: the Google Arts and Culture App allows cross-referencing between museums to enable the comparison of collections across different institutions. Immersive interactivity in museums and galleries is increasing, employing augmented reality, full-body effects such as responsive flooring, program mapping, effects such as fog and lasers and light-sculptures. Their blockbuster exhibitions such as *David Bowie Is* and Pink Floyd at the Victoria and Albert (V&A) Museum reflect pop-culture and mix objects with spectacle; the Alexander McQueen exhibition at the V&A in 2015 included a hologram of Kate Moss.

Fast authentication may be less affective – in fact the use of technology can be distancing. One of the authors visited the Terracotta Warriors exhibition at the Museum of Archaeology in Valletta, Malta, in March 2007. The exhibition simply featured the exhibits and was empty of other visitors, which meant that the encounter with the 2,000-year-old artefacts was more meaningful and, because the warriors are anthropomorphic, it was poignant to come face to face with them. In contrast, the Terracotta Warriors exhibition at the British Museum in 2007 was crowded, which interfered with participation and engagement (although, like all blockbuster exhibitions, timed tickets restricted the number of people able to attend). The range of interactive techniques in the exhibition was extensive, ranging from large-scale films to smaller effects, designed to authenticate the exhibition in magnified ways; however, they distracted the gaze. The spectacle conveyed the power and pomp of the original court of Qin Shi Huang and was evidently influenced by a wish to develop diplomatic relations with China, but the experience of the warriors had lost their immediacy and, like every theme park, required queuing to see each exhibit. While digitisation and special effects are prevalent forms of interpretation, natural exhibits – such as Sensational Butterflies at the Natural History Museum, allowing people to encounter live giant specimens – potentially create a stronger sense of aura, wonder, surprise and experiential authenticity than the animatronics in the Science Museum.

The final comparison with theme parks and their use of replication refers to the classic museum joke (author unknown), which shows a group of curators sitting around a table looking at a painting. The meeting chair says: 'Great! But how would it look on a fridge magnet?' Museum gift shops resemble theme parks in their range of merchandise, from the ubiquitous British Museum mummy pencil cases, designed as small-ticket items for school children, to the Ziggy Stardust album cover print for £2,250 at the V&A, illustrating how fake real and reproduction is increasingly becoming the business of museums.

Themed regions and places

The various themed places that have so far been considered have been, essentially, about themed attractions or concentrations of varied themed attractions. There are, however, other examples of theming that do not involve attractions, as such, but which relate to broader areas such as regions, counties and urban centres. Early examples of the theming of regions and counties can be found in the county and country *Shell Guides* (see also Chapter 7). The theming gives regional cloaks of meaning to places, which eventually permeates through into their identity.

The Shell Guide to Britain includes descriptions of the various counties and their sites of interest, with each one containing a montage-type painting containing key features associated with the county. Painted by famous artists, such as John Nash and Rowland Hilder, the pictures (or in some cases details from larger paintings) provide vivid symbols and images of the respective counties. The illustration for Staffordshire, for example, involving detail from an original painting by Julian Trevelyan, depicts a background of grimy furnaces surrounding the Potteries Museum, Stoke-on-Trent, 'as the symbol of the more refined products for which Staffordshire is famous' (Boumphrey, 1969, 480). The painting also contains a bust of Josiah Wedgewood, next to a blue jasper-ware vase and a mug of 'spit ware'; the Grand Trunk Canal, on which the county's prosperity was largely founded, runs through the centre of the picture. In addition to these particular features of Staffordshire, however, the county also possesses moorlands, part of the Peak District, to the north, and Cannock Chase, a mixed area of countryside and former royal forest, now designated as an Area of Outstanding Natural Beauty, in the south. Despite these attractive rural areas, however, the image portrayed in the picture is one of industrial heritage, which reinforces the Potteries theme and presents an impression that the county is synonymous with this particular image – an image that has been further reinforced through the novels of Arnold Bennett (see below). In contrast, the chapter covering Kent contains a detail from a painting by Rowland Hilder which includes attractive countryside, a picture of haymaking, hops and oast houses, blossom, an apple and a basket of cherries. This clearly reinforces the image of Kent as the 'Garden of England', a theme that is still used as a key strapline in marketing strategies by local tourist organisations to attract visitors to the county (see, e.g. Visit Kent and Visit South East England websites).

Broader regional images can also be shown to have been influenced through literature and film – for example, Thomas Hardy country (Dorset), Robin Hood country (Sherwood Forest) and Poldark country (Devon and Cornwall). The work of the 'Lake poets', especially Wordsworth, and the emphasis on romantic themes have been notable in creating place images of the Lake District (Robinson and Anderson, 2002), as have the stories of Beatrix Potter. Creative writing – for example, novels such as Gaskill's *North and South* – has also created a view of the North of England, although in this case the theme is a 'persistent image of darkness, depression, and dejection' (Robinson, 2002, p. 58). Shields (1991, cited in ibid., p. 59) identifies a stream of literary representation of grimness rooted in the nineteenth century, a tradition continued in the twentieth century in novels by authors such as Arnold Bennett, D.H. Lawrence and George Orwell. As Robinson (ibid.) points out, 'the tourist gaze toward the North is still clouded by a fictional legacy of mill smoke and drizzle'.

Similarly, Robert Burns and Sir Walter Scott, especially the latter, have had an influence on Scottish tourism, with Scott's works displaying his view of Scotland and Scottishness (ibid., p. 13). Queen Victoria also influenced the revival of many Scottish traditions and helped popularise them. Aitchison *et al.* (2000, p. 79) describe Scotland being 'endowed with an extremely rich iconography that is recognisable across the globe' with well-known icons including 'tartan, heather-clad hills, mist-strewn lochs, bagpipes, whisky, stags and a stalwart and colourful people that are often seen as a shorthand for the nation of Scotland as a whole'. They suggest that much of this enduring image is the result of literary and artistic representation and refer to the importance of Sir Walter Scott, whose novels and long poems encouraged the romance and whom Gold and Gold (1995, cited in ibid., p. 85) have described as a 'one-man Scottish Tourist Board'. The case of the Scottish Highlands, as Aitchison *et al.* (ibid., p. 79) point out,

> illustrates the close relationship between cultural representations of landscape and the creation of a resort for touristic consumption. The word 'myth' has been used to describe certain Highland/Scottish national and landscape ideals and values, and this word conveys much of the constructed and artificial nature of this land of the imagination.

The last word on invented Scottishness must go to *Outlander* (2015–), a television show which is mobilising the American filmic gaze. The curator of the West Highland Museum in Fort William had to read the novels, produce a leaflet on the subject and display the Jacobite artefacts, to feed demand of the Sassenachs.

Conclusion

This chapter has considered a variety of themed places, challenging the ideas that commerciality and replication is automatically inauthentic and demonstrating the complexity of fantastical reality. Some themed places are valued precisely because of their fake appearance and associated with playfulness and fun. Many of these places such as the various theme parks, the attractions of Las Vegas and the themed mega-malls constitute some of the most visited tourist destinations in the world. If places are memoirs, themed attractions are exaggerated, with their own untruthlike aura. While some theming may be largely accidental, resulting from the cultural assimilation of descriptions of places and attitudes expressed through film, television and literature, much of it is specifically constructed to enhance consumption, to attract more visitors and sell more merchandise. As discussed in Chapter 7, many places retain the image they had in their heyday that created an enduring imprint, which is then perpetuated.

Techniques of imagineered authenticity involve replication, simulacra and exaggerated reality, all of which could be said to be more accentuated in theme parks than in other environments. The truth is not simply varnished; the lily is well and truly gilded with extra bling. But far from being post-truth and post-authenticity, the imagineered authenticity in some of the parks, such as the Wizarding World of Harry Potter, adhere to two realities: the real world and the reality and rules of the fictional universe, which fans seek to mechanistically authenticate in a similar way to heritage sites. The undoubted attractiveness of such places to tourists would appear, at first sight, to cast doubt on the claims by MacCannell (1973) that tourists are always engaged in a quest for authenticity, suggesting perhaps a more postmodern view that they are sometimes engaged simply in a 'playful search for enjoyment' (Cohen, 1995). And yet, Pikkemaat and Schuckert (2007, p. 199) have argued that such attractions can fulfil tourists' needs for authenticity, as they would seem to be able to provide 'both personal fulfilment and escape to other places'. On one level, theme parks are designed as family venues; the meanings that people derive from them can be highly personal, constructing family myths and memories, as illustrated by Disney's Memory Maker photographic software.

Eco (1986, p. 44) posits that 'Disneyland tells us that faked nature corresponds much more to our daydream demands… [and] that technology can give us more reality than nature can.' The hyperreality of many themed places is vested in how they correspond more to idealised perceptions of what reality should be like than the real thing. In the Disney theme parks, everything is brighter, larger and more entertaining than in everyday life, with vinyl leaves on the trees (Fjellman, 1992). Like other chapters, this debate about themed places suggests that reproductions have a value. Rather than accepting faked reality as an inferior product, many visitors may prefer it and the type of experience it engenders. This idea forms part of the plot in a Julian Barnes (1998) novel, *England, England*, in which the Isle of Wight is transformed into a vast heritage centre containing replicas of everything 'English', from 'Buck House' to Stonehenge and Manchester United to the White Cliffs of Dover, based on the premise that replicas have their own significance.

Such places are 'both locationally and perceptually convenient' (Shaw and Williams, 2004, p. 255) and visitors may prefer clean, efficient and safe environments rather than the objective reality, which may not be as attractive. While some people – for example, as outlined in Chapter 8 – may seek out dangerous environments and be willing to take risks to achieve an 'adrenalin rush' or similar experience, it is likely that most people may not be so disposed. The immersion provided by a simulated earthquake or runaway train ride and a jungle with no stinging insects may be more pleasurable and thrilling than the real thing. Theme parks may seem to dictate and design 'fast authentication' for their visitors, they are theatres the spontaneous performances by human actors; however fake fake, fake real, or hyperreal the stage, experiences are co-created and people derive their own meaning from them.

Theme parks are archetypal real-fake places, which celebrate replicas, post-modern references and quotations. De-contextualisation is key to the fake-real place; as discussed earlier, Las Vegas places Venice beside Rome and Disneyland jumbles different eras and worlds together. They echo the Cast Courts at the V&A, which were provided as a method of surrogate travel for Victorians, where plaster copies of Trajan's Column are juxtaposed with the façade of Santiago di Compostela.

Today, Legoland recreates model places and compresses space still further, in their mini-meta version of 'Lego reality', which, in a further permutation, is threatened by fake counterfeit pieces on the market. The Lego versions of places are represented on postcards and fridge magnets and photographs, which are circulated as memes on social media, demonstrating that in Walter Benjamin's (1936 [2008]) 'Age of Mechanical Reproduction', replication is never-ending and that authenticity, like Russian dolls, offers infinite possibilities of copies of worlds within copies of worlds.

We end the book with a final variation on a theme, considering the example of the 'nested doll' or 'Matryoshka principle' applied to Michaelangelo's *David*. Firstly, the sculpture is a product of its influences: the inspiration of the Biblical figure, the classical statues on which it was based, the earlier *David* by Donatello and another by Verrocchio, the political will to create it and pride in the city-state, the choice of Michaelangelo as sculptor. *The David* was sculpted from a block of Carrara marble and (famously) disproportioned to be seen from the rooftop of Florence cathedral. In 1504 it was set on a plinth in the Palazzo della Signorina, looking in the direction of Rome. It was moved to the Accademia Gallery for safety (and concealed behind a brick wall during the Second World War) where it has required further protection from the reverberations of millions of tourist footsteps. A marble replica replaced it in the Palazzo and a bronze cast in Piazzale Michaelangelo is a tribute.

There are international reproductions - Cologne, Copenhagen, Marseilles, Mexico City – including a Victorian cast in the V&A Museum in London, which has a detachable fig leaf. This cast was scanned in 3-D in order to appear in the movie *Alien: Covenant*; the digital statue circulating in hyper-stylised tweets. An outsized work decorates a Queensland shopping mall and various misappropriations have ended up in court. Eco (1986) found an 'authentic copy' in front of the Palace of Living Arts in California and, more surprisingly, a wax version of the imagined artist's model inside. A black version of the statue titled *Noi* (Us) was created in solidarity with the victims of the Nice bombings. There is a Lego version. Art students model the veins of *David's* hands, touching their fingers to his fingers to the traces of Michaelangelo's fingers. Tourists emulate the singularity of the contrapposto stance in a photograph. We buy the mass-produced plaster souvenir, which gazes back at us from the shelf. We circulate the museum selfie meme on an iPad while Googling and discovering an Etsy soap carving imitation.

Those interested in mechanistic authenticity may know that one of the processes of casting bronze is called the 'cire perdue' ('lost wax'). The name refers to the formation of a clay mould around a wax figure. As the example of *David* suggests, authenticity concerns how wax (or truth) may melt away, but the shape it has made can be cast in continuous, infinitely emerging incarnations and possibilities.

References

Agarwal, S. (1994) 'The resort cycle revisited – implications for resorts', *Progress in Tourism, Recreation and Hospitality Management*, 5, pp. 194–207.

Agarwal, S. (1997) 'The public sector: planning for renewal?', in Shaw, G. and Williams A.M. (eds), *The Rise and Fall of British Coastal Resorts*, London: Pinter.

Agarwal, S. (1999) 'Restructuring and local economic development implications for seaside resort regeneration in Southwest Britain', *Tourism Management*, 20, pp. 511–22.

Agarwal, S. (2002) 'Restructuring seaside tourism: the resort lifecycle', *Annals of Tourism Research*, 29(1), pp. 25–55.

Agarwal, S. and Brunt, P. (2006) 'Social exclusion at English seaside resorts', *Tourism Management*, 27, pp. 654–70.

Agren, D. (2016) Mexico City's James Bond-inspired Day of the Dead parade gets mixed reviews, *Guardian*, https://www.theguardian.com/world/2016/oct/29/day-of-the-dead-parade-james-bond-mexico-city, accessed 10 November 2016, 10:00.

Aitchison, C., Macleod, N.E. and Shaw, S.J. (2000) *Leisure and Tourism Landscapes: Social and Cultural Geographies*, London: Routledge.

Alfonso, A.I., Kurti, L. and Pink, S. (eds) (2004) *Working Images: Visual Research and Representation in Ethnography*, London: Routledge.

ALVA (2016) *Visits Made in 2015 to Visitor Attractions in Membership with ALVA*, London: Association of Leading Visitor Attractions.

Appadurai, A. (1988) *The Social Life of Things: Commodities in Cultural Perspective*, Cambridge: Cambridge University Press.

Ashworth, G. (2008) 'In search of the place identity dividend: using heritage landscapes to create place identity', in Eyles, J. and Williams, A. (eds), *Sense of Place, Health and Quality of Life*, London: Routledge, pp.185–96.

Ashworth, G. and Tunbridge, J. (2000) *The Tourist-Historic City*, 2nd edn, London: Belhaven.

Ashworth, G.J., Graham, B. and Tunbridge, J.E. (2000) *A Geography of Heritage: Power, Culture and Economy*, London: Arnold.

Ashworth, G., Graham, B. and Tunbridge, J. (2007) *Pluralising Pasts: Heritage, Identity and Place in Multicultural Societies*, London: Pluto.

Askew, M. (2010) 'The magic list of global status', in Labadi, S. and Long, C., *Heritage and Globalisation*, Abingdon: Routledge, pp. 19–44.

Atkins, W. (2014) *The Moor: A Journey into the English Wilderness*. London: Faber and Faber.

Augé, M. (1995) *Non-places: Introduction to an Anthropology of Supermodernity*, London: Verso.

Australia Council (n.d.) http://www.australiacouncil.gov.au/aboriginal-and-torres-strait-islander-arts/, accessed 2 August 2016, 08:55.

Bachelard, G. (1994) *The Poetics of Space.* Massachusetts: Beacon Press

Bagnall, G. (1996) 'Consuming the past', in Edgell, S., Hetherington, K. and Warde, A. (eds), *Consumption Matters*, Oxford: Blackwell.

Baker, W.J. (1982) *Sports in the Western World*, Totowa, NJ: Rowman and Littlefield.

Bakhtin, M. (1984) *Rabelais and His World*, Bloomington: Indiana University Press.

Bakir, A. and Baxter, S.G. (2011) '"Touristic fun": motivational factors for visiting Legoland Windsor theme park', *Journal of Hospitality Marketing and Management*, 20(3–4), pp. 407–24.

Baldwin, E., Longhurst, B., McCraken, S., Ogborn, M. and Smith, G. (2004) *Introducing Cultural Studies*, Harlow: Pearson Education.

Barnes, J. (1998) *England, England*, London: Jonathon Cape.

Barson, T. (2016) 'O'Keeffe's century', in Barson, T. (ed.), *Georgia O'Keeffe*, London: Tate.

Barthes, R. (1979) *The Eiffel Tower and Other Mythologies*, London: University of California Press.

Barthes, R. (1980) *Camera Lucida*, London: Vintage.

Baskerville, S.W. and Willett, R. (1985) *Nothing Else to Fear: New Perspectives on America in the Thirties*, Manchester: Manchester University Press

Baudrillard, J. (1981) *Selected Writings: Simulacra and Simulation*, Cambridge: Polity Press.

Bayley, S. (2007) 'Is Bath Britain's most backward city', *Guardian*, https://www.theguardian.com/artanddesign/2007/sep/16/architecture, accessed 7 November 2016, 09:15.

BBC (2011) *Yellowstone: Unnatural History*, BBC Four Television Series, Season 1 – 16 June.

BBC (2016) *Imagine Georgia O'Keeffe: By Myself*, 26 July.

BBC News (2015) 'Ramblers launch app for "biggest-ever" footpath survey', www.bbc.co.uk/news/uk-33532041, accessed 28 April 2017.

BBC News (2016) 'Wanda steps up rivalry against Disney in China with second park', 26 September, www.bbc.co.uk/news/business-37469501, accessed 26 April 2017.

Beatty, C., Fothergill, S. and Wilson, I. (2011) *England's Smaller Seaside Towns: A Benchmarking Study*, Report by the Centre for Regional Economic and Social Research, Sheffield Hallam University, for the Department for Communities and Local Government.

Beedie, P. and Hudson, S. (2003) *Sport and Adventure Tourism*, New York: Haworth Press.

Beeton, S. (2005) *Film-induced Tourism*, Vol. 25, Bristol: Channel View.

Beeton, S. (2010) 'Landscapes as characters: film, tourism and a sense of place', *Metro Magazine: Media and Education Magazine*, 166, p. 114.

Beeton, S. (2015) *Travel, Tourism and the Moving Image*, Vol. 45, Bristol: Channel View.

Belhassen, Y. and Caton, K. (2006) 'Authenticity matters', *Annals of Tourism Research*, 33(3), pp. 853–6.

Beningfield, G. (1987) *Poems of the Countryside*, New York: Viking.

Beningfield, G. (1989) *Green and Pleasant Land: Poems of the Countryside*, New York: Viking.

Benjamin, W. (1936 [2008]) *The Work of Art in the Age of Mechanical Reproduction*, London: Penguin.

Benjamin, W. (1955) *Illuminations*, Glasgow: Fontana.

Benjamin, W. (2002) *The Arcades Project*, Belknap: Harvard University Press.

Benton, T. and Cecil, C. (2010) 'Heritage and public memory', in Benton, T. (ed.), *Understanding Heritage and Memory*, Manchester: Manchester University Press, pp. 7–43.

Benton, T. (2010) *Understanding Heritage and Memory*, Manchester: Manchester University Press.

Berger, A. (2004) *Deconstructing Travel: Cultural Perspectives on Tourism*, Lanham, MD: Rowman Altamira.

Berridge, G. (2007) *Events Design and Experience*, London: Routledge.

Best, R. and Shucksmith, M (2006) *Homes for Rural Communities*, Report of the Joseph Rowntree Foundation Rural Housing Policy Forum, York: Joseph Rowntree Foundation.

Bevolo, M. (2014) 'The dicourse of design as an asset for the city: from business innovation to vernacular event', in Richards, G., Marques, L. and Mein, K. (eds), *Event Design: Social Perspectives and Practices*, Abingdon: Routledge.

Bird, H. (2010) 'A personal view', in Bryson, B. (ed.), *Icons of England*, London: Transworld, pp. 76–9.

Birks, H.J.B. and Madsen, B.J. (1979) 'Flandrian vegetational history of Little Loch Roag, Isle of Lewis, Scotland', *Journal of Ecology*, 67, pp. 825–42.

Bishop, C. (2007) *Installation Art*, London: Routledge.

Bishop, C. (2012) *Artificial Hells: Participatory Art and the Politics of Spectatorship*, London: Verso.

Blyth, R. (1972) *Akenfield: Portrait of an English Village*, Harmondsworth: Penguin.

Boniface, B. and Cooper, C. (2009) *Worldwide Destinations: The Geography of Travel and Tourism*, 4th edn, Amsterdam: Elsevier Butterworth Heinemann.

Bonn, M., Joseph-Matthews, S., Dai, M., Hayes, S. and Cave, J. (2007) 'Heritage/cultural attraction atmospherics: creating the right environment for the heritage/cultural visitor', *Journal of Travel Research*, 45, pp. 345–54.

Boorstin, D. (1961 [1964]) *The Image: A Guide to Pseudo-Events in America*, New York: Harper & Row.

Booth, P. (1996) *The Role of Events in Glasgow's Urban Regeneration*, Strathclyde Papers on Planning No. 30, Glasgow: University of Strathclyde.

Bortolotto, C. (2010) 'Globalising intangible cultural heritage?', in Labadi, S. and Long, C., *Heritage and Globalisation*, Abingdon: Routledge, pp. 97–115.

Bossen, C. (2000) 'Festival mania, tourism, and nation-building in Fiji: the case of the Hibiscus Festival, 1956–1970', *The Contemporary Pacific*, 12(1).

Boumphrey, G. (1969) *The Shell Guide to Britain*, 2nd edn, London: Ebury Press.

Bourdieu, P. (1984) *Distinction*, London: Routledge.

Bowdin, G., Allen, J., O'Toole, W., Harris, R. and McDonnell, I. (2006) *Events Management*, 2nd edn, Oxford: Elsevier.

Bowdin, G., Allen, J., O'Toole, W., Harris, R. and McDonnell, I. (2011) *Events Management*, 3rd edn, Abingdon: Routledge.

Boyer, M. (1992) 'Cities for sale: merchandising history at South Street Seaport', in Sorkin, M. (ed.), *Variations on a Theme Park: The New American City and the End of Public Space*, New York: Hill and Wang, pp. 181–204.

Boyer, M. (1996) *The City of Collective Memory: Its Historical Imagery and Architectural Entertainments*, Cambridge, MA: MIT Press.

Boyle, M. and Hughes, G. (1994) 'The politics of urban entrepreneurialism in Glasgow', *Geoforum*, 25(4), pp. 453–70.

Breakell, M. and Human, B. (1999) 'Practical guidance for tourism management in historic towns: making the connections', Conference proceedings, Oxford Brooks University, Oxford, English Historic Towns Forum (EHTF).

Brodie, A. and Winter, G. (2007) *England's Seaside Resorts*, Swindon: English Heritage.

Brown, C., Mcmorran, R. and Price, M.F. (2011) 'Rewilding: a new paradigm for nature conservation in Scotland', *Scottish Geographical Journal*, 127(4), pp. 288–314.

Brown, D. (1996) 'Genuine fakes', in Selwyn, T. (ed.), *The Tourist Image: Myths and Myth Making in Tourism*, Chichester: Wiley, pp. 33–47.

Bruner, E.M. (1991) 'Transformation of self in tourism', *Annals of tourism Research*, 18(2), pp. 238–50.

Bruner, E. (1994) 'Abraham Lincoln as authentic reproduction: a critique of postmodernism', *American Anthropologist*, 96, pp. 397–415.

Bryman, A. (2004) *The Disneyization of Society*, London: Sage.

Bryman, A. (2009) 'Global implications of McDonaldization and Disneyization', in Ritzer, G., *McDonaldization: The Reader*, 3rd edn, London: Sage, pp. 395–410.

Bryson, B. (2015) *The Road to Little Dribbling: More Notes from a Small Island*, London: Doubleday.

Bull, C. and Bonser, R. (1988) *The Recreational Use of Priory Country Park: A Visitor Survey*, Report by the Conservation and Recreation Research Unit (Bedford College of Higher Education) for North Bedfordshire Borough Council's Leisure and Tourism Department.

Bull, C.J. and Hayler, S. (2009) 'The changing role of live entertainment at English seaside resorts at the beginning of the 21st century', *Tourism Geographies*, 11, pp. 281–307.

Bull, C. and Lovell, J. (2007) 'The impact of hosting mega sporting events on local residents: an analysis of the views and perceptions of Canterbury residents in relation to the *Tour de France* 2007', *Journal of Sport and Tourism*, 12(3–4), pp. 229–48.

Bull, S. (2008) 'Street and studio' lecture, Tate Modern, 5 August.

Bunce, M. (1994) *The Countryside Ideal: Anglo-American Images of Landscape*, London: Routledge.

Burns, J.P.A., Hatch, J.H. and Mules, T.J. (1986) *The Adelaide Grand Prix: The Impact of a Special Event*, Adelaide: Centre for South Australian Economic Studies.

Burrell, I. (2016) 'ITV's head of drama Steve November on *Victoria*, the new series hoping to rival *Downton Abbey*', *Independent*, http://www.independent.co.uk/arts-entertainment/tv/news/itvs-head-of-drama-steve-november-on-victoria-the-new-series-hoping-to-rival-downton-abbey-a6844261.html, accessed 9 December 2016.

Butcher, J. (2006) 'Cultural politics, cultural policy and cultural tourism', in Smith, M. and Robinson, M., *Cultural Tourism in a Changing World*, Bristol: Channel View, pp. 21–35.

Butler, R.W. (1980) 'The concept of a tourist area cycle of evolution: implications for management of resources', *Canadian Geographer*, 24, pp. 5–12.

Butler, R. (1985) 'Evolution of tourism in the Scottish Highlands', *Annals of Tourism Research*, 12, pp. 371–91.

Butler, R. (1991) 'West Edmonton Mall as a tourist attraction', *Canadian Geographer*, 35, pp. 287–95.

Butler, R.W., Hall, C.M. and Jenkins, J. (eds) (1998) *Tourism and Recreation in Rural Areas*, Chichester: Wiley.

CABE (2003) *Shifting Sands*, English Heritage, http://webarchive.nationalarchives.gov.uk/20110118095356/http:/www.cabe.org.uk/files/shifting-sands.pdf, accessed 28 January 2017, 17:00.

Calais-guide (n.d.) http://www.calais-guide.co.uk/, accessed 19 April 2017, 08:50.

Canjels, R. (2011) *Distributing Silent Film Serials: Local Practices, Changing Forms, Cultural Transformation*, London: Routledge.

Carnegie, E. and McCabe, S. (2008) 'Re-enactment events and tourism: meaning, authenticity and identity', *Current Issues in Tourism*, 11(4), pp. 349–68.

Cassidy, F. (2006) 'What motivates event tourists?', in *Proceedings of Business and Information*, Academy of Taiwan Information Systems Research, international conference on business and information, 12–14 July, Singapore.

Chakrabortty, A. (2011) 'Why doesn't Britain make things any more?', *Guardian*, http://www.guardian.co.uk/business/2011/nov/16/why-britain-doesnt-make-things-manufacturing, accessed 1 January 2012, 17:10.

Chalip, L. (2006) 'Towards social leverage of sport events', *Journal of Sport and Tourism*, 11(2), pp. 109–27.

Chambers, D. and Rakić, T. (2012) *An Introduction to Visual Research Methods in Tourism*, Abingdon: Routledge.

Chhabra, D. (2005) 'Defining authenticity and its determinants: toward an authenticity flow model', *Journal of Travel Research*, 44(1), pp. 64–73.

Chronis, A. (2006) 'Heritage of the senses: collective remembering as embodied praxis', *Tourist Studies*, 6, pp. 267–96.

Clarke, D. (2003) *The Consumer Society and the Postmodern City*, Abingdon: Routledge.

Clarke, D. (2005) *The Cinematic City*, London: Routledge.

Clout, H. (1981) 'Rural settlements', *Progress in Human Geography*, 5, pp. 408–13.

Coates, N. (2000) 'Brief encounters', in Miles, M., Hall, T. and Borden, I. (eds), *The City Cultures Reader*, London: Routledge.

Cobbett, W. ([1830] 1967) *Rural Rides*, Harmondsworth: Penguin.

Cohen, E. (1979a) 'Rethinking the sociology of tourism', *Annals of Tourism Research*, 6(1), pp. 18–35.

Cohen, E. (1979b [1996]) 'A phenomenology of tourist experiences', originally published in *Sociology*, 13, pp. 179–201, in Apostolopoulos, Y. and Yiannais, A., *The Sociology of Tourism: Theoretical and Empirical Investigations*, London: Routledge, pp. 90–114.

Cohen, E. (1988a) 'Authenticity and commoditization in tourism', *Annals of Tourism Research*, 15, pp. 371–86.

Cohen, E. (1988b) 'Traditions in the sociology of tourism', *Annals of Tourism Research*, 15, pp. 29–46.

Cohen, E. (1995) 'Contemporary tourism trends and challenges: sustainable authenticity or contrived post-modernity?', in Butler, R. and Pearce, D. (eds), *Change in Tourism: People, Places, Processes*, London: Routledge, pp. 12–29.

Cohen, E. (1997) 'The iconography of the tourism experience', *Annals of Tourism Research*, 24(4), pp. 951–69.

Cohen, E. (2002) 'Authenticity, equity and sustainability in tourism', *Journal of Sustainable Tourism*, 10(4), pp. 267–76.

Cohen, E. (2007) '"Authenticity" in tourism studies: aprés la lutte', *Tourism Recreation Research*, 32(2), pp. 75–82.

Cohen, E. (2012) 'The dilemma of authenticity and inauthenticity', in Singh, T., *Critical Debates in Tourism*, Bristol: Channel View, pp. 250–61.

Cohen, E. and Cohen, S.A. (2012) 'Authentication: hot and cool', *Annals of Tourism Research*, 39(3), pp. 1295–314.

Coleman, S. (2008) 'Pilgrimage, memory and the materialisation of the past at Walsingham', Paper delivered at Canterbury Christ Church University Conference, 'Dimensions of Pilgrimage: Journey, Meaning and Place', 13 September.

Coleman, S. and Crang, M. (2002) 'Grounded tourists, travelling theory', in Coleman, S. and Crang, M. (eds), *Tourism: Between Place and Performance*, Oxford: Berghahn, pp. 1–20.

Coleman, S. and Eade, J. (2004) *Reframing Pilgrimage: Cultures in Motion*, Abingdon: Psychology Press.

Collins, M.F. and Jackson, G.A.M. (1999) 'The economic impact of sport and tourism', in Standeven, J. and De Knop, P., *Sport Tourism*, London: Human Kinetics.

Collins-Kreiner, N. (2010) 'Researching pilgrimage: continuity and transformations', *Annals of Tourism Research*, 37(2), pp. 440–56.

Comunian, R. and Mould, O. (2014) 'The weakest link: creative industries, flagship cultural projects and regeneration', *City, Culture and Society*, 5(2), pp. 65–74.

Conisbee, M., Kjell, P., Oram, J., Bridges-Pslmer, J., Simms, A. and Taylor, J. (2004) *Clone Town Britain: The Loss of Identity on the Nation's High Streets*, London: New Economics Foundation.

Connerton, P. (1989) *How Societies Remember*, London: Cambridge University Press.

Cooper, C.P. (1990) 'Resorts in decline: the management response', *Tourism Management*, 11(1), pp. 63–7.

Cooper, C (1992) 'The life cycle concept and strategic planning for coast resorts', *Built Environment*, 18(1), pp. 57–66.

Cooper, C. (1997) 'Parameters and indicators of the decline of the British seaside resort', in Shaw, G. and Williams, A.M. (eds), *The Rise and Fall of British Coastal Resorts*, London: Pinter.

Cooper, C. and Jackson, S. (1989) 'Destination life cycle: the Isle of Man case study', *Annals of Tourism Research*, 16, pp. 377–98.

Coppock, J.T. (1977) 'The challenge of change: problems of rural land use in Britain', *Geography*, 62(2), p. 78.

Cosgrove, D. (1984) *Social Formation and Symbolic Landscape*, London: Croom Helm.

Cosgrove, I. and Jackson, R. (1972) *The Geography of Recreation and Leisure*, London: Hutchinson.

Coslett, P. (2012) Sea Odyssey giants capture Liverpool's heart, BBC, http://www.bbc.co.uk/news/uk-england-merseyside-17806029, accessed 4 December 2016, 18:45.

Countryside Commission (1988) *National Rights of Way Condition Survey 1988*, Cheltenham: Countryside Commission.

Coverley, M. (2006) *Psychogeography*, Harpenden: Pocket Essentials.

Crang, M. (1996) 'Magic kingdom or a Quixotic quest for authenticity?', *Annals of Tourism Research*, 23(2), pp. 415–31.

Crang, M. (2006) 'Circulation and emplacement: the hollowed out performance of tourism', in Minca, C. and Oakes, T. (eds), *Travels in Paradox: Remapping Tourism*, Lanham, MD: Rowman & Littlefield, pp. 47–64.

Crang, M. and Thrift, N. (eds) (2000) *Thinking Space*, London: Routledge.

Crang, P. (1997) 'Performing the tourist product', in Rojek, C. and Urry, J. (eds), *Touring Cultures: Transformations of Travel and Theory*, London: Routledge, pp. 137–54.

Crawford, M. (1992) 'The world in a shopping mall', in Sorkin, M. (ed.), *Variations on a Theme Park: The New American City and the End of Public Space*, New York: Hill and Wang, pp. 3–30.

Cresswell, T. (2006) *Place: A Short Introduction*, Oxford: Blackwell.

Crewe, L. and Beaverstock, J. (1998) 'Fashioning the city: cultures of consumption in contemporary urban spaces', *Geoforum*, 29(3), pp. 287–308.

Cronon, W. (2003) 'The trouble with wilderness, or, getting back to the wrong nature', in Warren, L.S. (ed.), *American Environmental History*, Oxford: Blackwell.

Crouch, D. (2000) 'Places around us: embodied lay geographies in leisure and tourism', *Leisure Studies*, 19, pp. 63–77.

Crouch, D. (2002) 'Surrounded by place: embodied encounters', in Coleman, S. and Crang, M. (eds), *Tourism: Between Place and Performance*, Oxford: Berghahn, pp. 207–18.

Crouch, D. and Lübbren, N. (eds) (2003) *Visual Culture and Tourism*, Oxford: Berg.

Croy, G. (2004) *The Lord of the Rings, New Zealand and Tourism: Image Building with Film*, Department of Management Working Paper Series, Working Paper 10/04, Monash University.

Csíkszentmihályi, M. and Csíkszentmihályi, I.S. (eds) (1988) *Optimal Experience: Psychological Studies of Flow in Consciousness*, New York: Cambridge University Press, pp. xiv, 364–83.

Cudny, W. (2016) *Festivalisation of Urban Spaces: Factors, Processes and Effects*, New York: Springer.

Cullen, S.M. (2016) *In Search of the Real Dad's Army: The Home Guard and the Defence of the United Kingdom*, Barnsley: Pen & Sword Military.

Curry, N. (1994) *Countryside Recreation, Access and Land Use Planning*, London: Spon.

Davidson, M. (2015) 'Best places to commute from the seaside', *Telegraph*, 13 March.

Davies, N. (1997) *Europe: A History*, London: Pimlico.

DCMS (2016) *The Culture White Paper* https://www.gov.uk/government/uploads/system/uploads/attachment_data/file/510798/DCMS_The_Culture_White_Paper__3_.pdf, accessed 18 October 2016.

De Certeau, M. (1984) *The Practice of Everyday Life*, Berkeley: University of California Press.

Debord, G. (1983 [1987]) *The Society of the Spectacle*, London: Rebel Press.

DEFRA (2006) *England Leisure Visits: Report of the 2005 Survey*, Bristol: Department for Environment, Food and Rural Affairs.

DEFRA (2015) *Towards a One Nation Economy: A 10-Point Plan for Boosting Productivity in Rural Areas*, Bristol: Department for Environment, Food and Rural Affairs.

Delyser, D. (1999) 'Authenticity on the ground: engaging the past in a California ghost town', *Annals of the Association of American Geographers*, 89(4), pp. 602–32.

Demetriadi, J. (1997) 'The golden years: English seaside resorts 1950–1974', in Shaw, G. and Williams A.M. (eds), *The Rise and Fall of British Coastal Resorts*, London: Pinter.

Department of Culture, Media and Sport (1999) *Tomorrow's Tourism*, London: DCMS.

Department of Culture, Media and Sport (2008) *Sea Changes*, London: DCMS.

Department for Communities and Local Government (2008) *England's Seaside Towns: A 'Benchmarking Study'*, London: DCLG.

Department for Communities and Local Government (2010) *Strategy for Seaside Success: Securing the Future of Seaside Economies*, London: DCLG.

Department for Communities and Local Government (2011) *English Indices of Deprivation 2010*, London: DCLG.

Desbordes, M. (2007) 'A review of the economic impact studies done on the Tour de France: methodological aspects and first results', *International Journal of Sport Management and Marketing*, 2(5/6), pp. 526–40.

Dimbleby, J. (2010) 'Beside the seaside', in Bryson, B. (ed.), *Icons of England*, London: Transworld, pp. 139–41.

Donald, J. (1992) 'Metropolis: the city as text', in Bocock, R. and Thompson, K. (eds), *Social and Cultural Forms of Modernity*, Cambridge: Open University, pp. 417–61.

Donald, J. (1999) *Imagining the Modern City*, Minneapolis: University of Minnesota Press.

Drabble, M. (1979) *A Writer's Britain: Landscape in Literature*, London: Thames and Hudson.
Dryburgh, L. (2010) 'Culture-led regeneration in seaside towns', in Walton, J. and Browne, P. (eds), *Coastal Regeneration in English Resorts – 2010*, Lincoln: Coastal Communities Alliance.
Dudley, N. (2011) *Authenticity in Nature*, London: Earthspan.
Ebner, F. (2008) 'Urban characters, imaginary cities', in Eskildsen, U., *Street and Studio: An Urban History of Photography*, London: Tate Publishing, pp. 186–93.
Eco, U. (1986) *Faith in Fakes: Travels in Hyperreality*, London: Picador.
Edensor, T. (2001) 'Performing tourism, staging tourism: (re) producing tourist space and practice', *Tourist Studies*, 1, pp. 59–80.
Edensor, T. (2010) 'Walking in rhythms: place, regulation, style and the flow of experience', *Visual Studies*, 25(1), pp. 69–79.
Edmonton Journal (2015) 'West Edmonton Mall Facts', 2 February.
Edwards, E. (1996) 'Greeting from another world', in Selwyn, T. (ed.), *The Tourist Image: Myths and Myth Making in Tourism*, Chichester: Wiley, pp. 197–221.
Edwards, S. (2006) *Photography: A Very Short Introduction*, Oxford: Oxford University Press.
Edwards, S. (2008) Lecture: 'Street and Studio', Tate Modern, 5 July.
Ehrenreich, B. (2007) *Dancing in the Streets: A History of Collective Joy*, London: Granta.
Eliasson O. (2016a) http://olafureliasson.net/archive/artwork/WEK110140/waterfall, accessed 1 August 2016, 09:00.
Eliasson, O. (2016b) http://www.publicartfund.org/view/exhibitions/5717_the_new_york_city_waterfalls, accessed 1 August 2016, 16:45.
Emmison, M. and Smith P. (2000) *Researching the Visual*, London: Sage.
English Tourist Board (1984) *Curtain Up on the Resorts*, London: English Tourist Board.
English Tourism Council (2001) *Sea Changes: Creating World-Class Resorts in England*, London: ETC.
Ensamble Studio (n.d.) http://www.ensamble.info/#!tippet-rise-art-center/i9j0o, accessed 1 August 2016, 19:45.
Essex, S. and Chalkley, B. (1998) 'Olympic Games: catalyst of urban change', *Leisure Studies*, 17(3), pp. 187–206.
European Association of Historic Towns and Regions (2002) Constitution.
Evans and Shaw (2004) *The Contribution of Culture to Regeneration in the UK: A Review of Evidence*, Report to the Department for Culture, Media and Sport, London: London Metropolitan University.
Everett, S. (2012) 'Production places or consumption spaces? The place-making agency of food tourism in Ireland and Scotland', *Tourism Geographies*, 14(4), pp. 535–54.
Fairclough, G. (2008) 'New heritage, an introductory essay: people, landscape, change', in Fairclough, G., Harrison, R., Jameson, J. and Schofield, J. (eds), *The Heritage Reader*, Abingdon: Routledge, pp. 297–312.
Fairclough, G., Harrison, R., Jameson, J. and Schofield, J. (eds) (2008) *The Heritage Reader*, Abingdon: Routledge.
Falk, P. and Campbell, C. (eds) (1997) *The Shopping Experience*, London: Sage.
Faris, W.B. (2004) *Ordinary Enchantments: Magical Realism and the Remystification of Narrative*, Nashville, TN: Vanderbilt University Press.
Farley, P. and Simmons Roberts, M. (2011) *Edgelands: Journeys into England's True Wilderness*, London: Jonathon Cape.

Ferdinand, N. and Shaw, S.J. (2012). 'Events in our changing world', in Ferdinand, N. and Kitchen, P. (eds), *Events Management: An International Approach*. London: Sage.

Ferry, K. (2009) *The British Seaside Holiday*, Oxford: Shire.

Finley, M.I. and Pleket, H.W. (1976) *The Olympic Games*, Edinburgh: R. and R. Clark.

Finkel, R. (2010). '"Dancing around the ring of fire": social capital, tourism resistance, and gender dichotomies at up Helly Aa in Lerwick, Shetland', *Event Management*, 14(4), pp. 275–85.

Finn, A. and Erdem, T. (1995) 'The economic impact of a mega-multi mall: estimation issues in the case of West Edmonton Mall', *Tourism Management*, 16(5), pp. 367–73.

Fisher, P. (1998) *Wonder, the Rainbow, and the Aesthetics of Rare Experiences*, Cambridge, MA: Harvard University Press.

Fitch, J.M. (1990) 'Vernacular paradigms for post-industrial architecture', *Vernacular Architecture*, Aldershot: Avebury, pp. 261–70.

Fjellman, S.M. (1992) *Vinyl Leaves: Walt Disney World and America*, Boulder, CO: Westview Press.

Fletcher, A. and Lee, M.J. (2012) 'Current social media uses and evaluations in American museums', *Museum Management and Curatorship*, 27(5), pp. 505–21.

Flew, T. (2012) *The Creative Industries: Culture and Policy*, London: Sage.

Florida, R. (2002) *The Rise of the Creative Class: And How It's Transforming Work, Leisure, Community and Everyday Life*, New York: Basic.

Foley, M. and McPherson, G. (2007) 'Glasgow's Winter Festival: can cultural leadership serve the common good?', *Managing Leisure*, 12(2–3), pp. 142–56.

Foley, M., McPherson, G. and Matheson, C. (2006) 'Glocalisation and Singaporean festivals', *International Journal of Event Management Research*, 1(2), pp. 1–16.

Foley, M., McPherson, G. and Matheson, C. (2007) 'Cultural identity and festivity: generating Singapore through citizenship and enterprise in events activity', in Aitchison, C. and Pritchard, A. (eds), *Festivals and Events: Culture and Identity in Leisure, Sport and Tourism*, LSA Publication No. 94, Brighton.

Foley, M., McGillivray, D. and McPherson, G. (2009) 'Establishing Singapore as the events and entertainment capital of Asia: strategic brand diversification', in Ali-Knight, J., Robertson, M., Fyall, A. and Ladkin, A. (eds), *International Perspectives of Festivals and Events, Paradigms of Analysis*, London: Elsevier.

Foley, M., McGillivray, D. and McPherson, G. (2012) *Event Policy: From Theory to Strategy*, London: Routledge.

Foucault, M. (1984) 'Of other spaces: utopias and heterotopias' (1967), *Architecture/Mouvement/Continuite*, October, trans. Miskowiec, J.

Foucault, M. (1995) *Discipline and Punish: The Birth of the Prison*, New York: Knopf Doubleday.

Foucault, M. and Miskowiec, J. (1986) 'Of other spaces', *diacritics*, 16(1), pp. 22–7.

Fox, K. and Lovell, J. (2015) 'Landscape without borders: an analysis of the role of the desert in the narrative seriality and visual storytelling of Breaking Bad 2008–2013', Breaking Borders Humanities Conference, Canterbury Christ Church University.

Franklin, A. (2003) *Tourism: An Introduction*, London: Sage.

Frayling C. (2007) in Bayley, S., 'Is Bath England's most backward city?', *Guardian* https://www.theguardian.com/artanddesign/2007/sep/16/architecture, accessed 28 October 2016, 9:40.

Freeman, S. (2015) *Beside the Sea: Britain's Lost Seaside Heritage*, London: Aurum Press.

Frost, W. and Laing, J. (2015) *Imagining the American West Through Film and Tourism*, London: Routledge.

Gale, T. (2005) 'Modernism, post-modernism and the decline of British seaside resorts as long holiday destinations: a case study of Rhyl, North Wales', *Tourism Geographies*, 7(1), pp. 86–112.

Gallent, N., Iqbal, H., Juntti, M., Kidd, U. and Shaw, D. (2015) *Introduction to Rural Planning: Economics, Communities and Landscapes*, London: Routledge.

Galliver, J. (2013) http://www.bbc.co.uk/news/magazine-22467510, BBC Real Time interviews, accessed 7 January 2016, 09:00.

Garcia, B. (2005) 'Deconstructing, the city of culture; the long-term cultural legacy of Glasgow, 1990', *Urban Studies*, 42(5/6).

Garrod, B. (2008) 'Exploring place perception: a photo-based analysis', *Annals of Tourism Research*, 35(2), pp. 381–401.

Getz, D. (1997) *Event Management and Event Tourism*, 1st edn, New York: Cognizant.

Getz, D. (2005) *Festivals, Special Events and Tourism*, 2nd edn, New York: Van Nostrand.

Getz, D. (2007) *Event Studies: Theory, Research and Policy for Planned Events*, Oxford: Butterworth Heinemann.

Getz, D. and Page, S. (2010) *Event Studies: Theory, Research and Policy for Planned Events*, London: Routledge.

Gilbert, D. and Lizotte, M. (1998) 'Tourism and the performing arts', *Travel and Tourism Analyst*, 1, pp. 82–96.

Gilg, A. (1996) *Countryside Planning: The First Half Century*, 2nd edn, London: Routledge.

Gilmore, A. (2016) *Fantastic Beasts and J.K. Rowling*, ITV, 27 December 2016, 17:40.

Glyptis, S. (1991) *Countryside Recreation*, Harlow: Longman/ILAMimpacts.

Goffman, E. (1959) *The Presentation of Self in Everyday Life*, New York: Doubleday.

Gold, J.R. (1980) *An Introduction to Behavioural Geography*, Oxford: Oxford University Press.

Gold, J. and Gold, M. (1995) *Imagining Scotland: Tradition, Representation, and Promotion in Scottish Tourism since 1750*, Aldershot: Scolar.

Goodall, B. (1993) 'Industrial heritage and tourism', *Built Environment*, 19(2), p. 93.

Goodey, B. (2006) 'Interpreting urban heritage', in Hems, A. and Blockley, M., *Heritage Interpretation*, Abingdon: Routledge, pp. 9–32.

Goss, J. (1993) '"The Magic of the Mall": an analysis of form, function and meaning in the contemporary retail built environment', *Annals of the Association of American Geographers*, 83, pp. 18–47.

Gottdiener, M. (1997 [2001]) *The Theming of America: American Dreams, Media Fantasies and Themed Environments*, Boulder, CO: Westview.

Gottdiener, M. (1982) 'Disneyland: a utopian urban space', *Journal of Contemporary Ethnography*, 11(2), pp. 139–62.

Gottdeiner, M. (2001) *Life in the Air*, Oxford: Rowman and Littlefield.

Graburn, N. (1989) 'Tourism: the sacred journey', in Smith, V., *Hosts and Guests: The Anthropology of Tourism*, Philadelphia: University of Pennsylvania Press, pp. 21–36.

Grant, M., Human, B. and le Pelley, B. (1999a) 'Heritage and making the most of it', Tourism Intelligence Papers, *Insights*, London: ETC, September, pp. A37–42.

Grant, M., Human, B. and le Pelley, B. (1999b) 'Making the connections: joined up thinking in action', Tourism Intelligence Papers, *Insights*, London: ETC, July, pp. A20–3.

Grant, M., Human, B. and le Pelley, B. (2002) 'Embracing tourism: Cambridge tourism strategy review case study', *Insights*, London: ETC, March, pp. C41–50.

Gratton, C., Shibli, S. and Coleman, R. (2005) 'The economics of sport tourism at major sports events', in Higham, J. (ed.), *Sport Tourism Destinations: Issues, Opportunities and Analysis*, Oxford: Elsevier.

Gravari-Barbas, M. and Graburn, N. (eds) (2016) *Tourism Imaginaries at the Disciplinary Crossroads: Place, Practice, Media*, London: Routledge.

Gray, A. and Boyd, W. (2007) *Lanark: A Life in Four Books*, Vol. 100, first published 1981, Edinburgh: Canongate.

Gray, F. (2006) *Designing the Seaside: Architecture, Society and Nature*, London: Reaktion.

Gray, F. and Powell, D. (2010) 'Cultural regeneration in four South East coastal towns', in Walton, J. and Browne, P. (eds), *Coastal Regeneration in English Resorts – 2010*, Lincoln: Coastal Communities Alliance.

Great Britain Day Visits Survey (2015) *The GB Day Visitor: Statistics 2015*, Edinburgh: TNS.

Great Britain Tourism Survey (2015) *GB Tourism Survey: 2015 Overview*, London: Visit Britain.

Haldrup, M. and Larsen, J. (2003) 'The family gaze', *Tourist Studies*, 3, pp. 23–45.

Haldrup, M. and Larsen, J. (2012) 'Readings of tourist photographs', in Chambers, D. and Rakić, T., *An Introduction to Visual Research Methods in Tourism*, Abingdon: Routledge, pp. 153–68.

Halewood, C. and Hannam, K. (2001) 'Viking heritage tourism: authenticity and commodification', *Annals of Tourism Research*, 28(3), pp. 565–80.

Hall, C.M. (2005) 'Reconsidering the geography of tourism and contemporary mobility', *Geographical Research*, 43(2), pp. 125–39.

Hall, C. (2012) 'Sustainable mega-events: beyond the myth of balanced approaches to mega-event sustainability', *Event Management*, 16(2), pp. 119–31.

Hall, C.M. and Page, S. (2006) *The Geography of Tourism and Recreation: Environment, Place and Space*, 3rd edn, London: Routledge.

Hall, M. (1992) 'Adventure, sport and health tourism' in Weiler B and Hall, M (eds) *Special Interest Tourism*, London: Belhaven.

Hall, P.G. (1973) *The Containment of Urban England: The Planning System: Objectives, Operations*, London: Allen and Unwin.

Hall, S. (2008) 'Whose heritage? Un-settling "The Heritage", re-imagining the post-nation', in Fairclough, G., Harrison, R., Jameson, J. and Schofield, J. (eds), *The Heritage Reader*, Abingdon: Routledge, pp. 219–28.

Hannam, K. and Halewood, C. (2006) 'European Viking themed festivals: an expression of identity', *Journal of Heritage Tourism*, 1(1), pp. 17–31.

Hannam, K. and Knox, D. (2010) *Understanding Tourism: A Critical Introduction*, London: Sage.

Hannigan, J. (1998) *Fantasy City: Pleasure and Profit in the Postmodern Metropolis*, London: Routledge.

Hannigan, J. (2007) 'Casino cities', *Geography Compass*, 1(4), pp. 959–75.

Harper, D. (1998) 'An argument for visual sociology', in Prosser, J. *Image-based Research: A Sourcebook for Qualitative Researchers*, Abingdon: Psychology Press, pp. 24–41.

Harper, G. and Rayner, J. (2010) 'Introduction: cinema and landscape', *Cinema and Landscape*, Bristol: Intellect, pp. 15–28.

Harrison, C. (1991) *Countryside Recreation in a Changing Society*, London: TMS.

Harrison, R. (1994) *Manual of Heritage Management*, Oxford: Butterworth Heinemann.

Harrison, R. (2010) 'What is heritage?', in Harrison, R. (ed.), *Understanding the Politics of Heritage*, Manchester: Manchester University Press, pp. 154–96.

Harrison, R. (2013) *Heritage: Critical Approaches*, Abingdon: Routledge.

Harrison, R. and Hitchcock, M. (2010) 'The politics of heritage', Harrison, R. (ed.), *Understanding the Politics of Heritage*, Manchester: Manchester University Press, pp. 154–96.

Harrison R. and Rose D. (2010) 'Intangible heritage', in Benton, T. *Understanding Heritage and Memory*, Manchester: Manchester University Press, pp. 238–76.

Harrison, R., Jameson, J. and Schofield, J. (eds) (2008) 'Heritage, memory and modernity: an introduction', *The Heritage Reader*, Abingdon: Routledge.

Harvey, D. (1989a) *The Urban Experience*, Baltimore, MD: Johns Hopkins University Press.

Harvey, D. (1989b) *The Condition of Postmodernity*, Oxford: Basil Blackwell.

Hayler, S.M. (1999) 'Live entertainment at the seaside', *Cultural Trends*, 34, pp. 63–98.

Hayler, S.M. (2004) 'An analysis of the role of live entertainment at English seaside resorts, with particular reference to the 1990s and the beginning of the 21st century', unpublished doctoral thesis, University of Kent.

Hayllar B. and Griffin, T. (2005) 'The precinct experience: a phenomenological approach', *Tourism Management*, 26, pp. 517–28.

Hayllar, B., Griffin, T. and Edwards, D. (eds) (2008) *City Spaces, Tourist Places: Urban Tourism Precincts*, Oxford: Butterworth Heinmann.

Haywood, K.M. (1986) 'Can the tourist area life cycle be made operational?', *Tourism Management*, 7, pp. 154–67.

Heathcote, D. (2011) *A Shell Eye on England: The Shell County Guides 1934–1984*, Faringdon: Libri.

Heidegger, M. (1971) *Poetry, Language, Thought*, New York: HarperCollins.

Henkel, L.A. (2014) 'Point-and-shoot memories the influence of taking photos on memory for a museum tour', *Psychological Science*, 25(2), pp. 396–402.

Henry, N. and Massey, D. (1995) 'Competitive time-space in high technology', *Geoforum*, 26(1), pp. 49–64.

Hern, A. (1967) *The Seaside Holiday*, London: Cresset Press.

Heron, E.L. (2004) 'Placing geographical imagination in film: New Zealand filmmakers' use of landscape', *New Zealand Geographer*, 60(1), pp. 60–6.

Hewison, R. (1987) *The Heritage Industry: Britain in a State of Decline*, London: Methuen.

Hirschman, E. and Holbrook, M. (1982) 'Hedonic consumption: emerging concepts, methods and propositions', *Journal of Marketing*, 46(3), pp. 92–101.

Historic England (2015) Heritage Counts Report, https://content.historicengland.org.uk/content/heritage-counts/pub/2016/placemaking-and-heritage-infographic.pdf, accessed 8 January 2016.

Hobsbawm, E. and Ranger, T. (eds) (1983) *The Invention of Tradition*, Oxford: Blackwell.

Hollinshead, K. (1998) 'Disney and commodity aesthetics: a critique of Fjellman's analysis of "Distory" and the "Historicide" of the past', *Current Issues in Tourism*, 1(1), pp. 58–119.

Hooper, M.D. (1977) 'Hedgerows and small woodlands', in Lloyd, R.J. and Davidson, J. (eds), *Conservation and Agriculture*, Chichester: John Wiley, p. 49.

Howell, F.C. (2013). 'Sense of place, heterotopia, and community: performing land and folding time in the Badalisc Festival of Northern Italy', *Folklore*, 124(1), pp. 45–63.

Hudd, D. (2016) 'The future for UK arts, culture and heritage: implementing the new strategy', Westminster Media Forum, 7 June.

Hudson, S. (2003) *Sport and Adventure Tourism*, New York: Haworth Hospitality Press.

Hudson, S. and Ritchie, J.B. (2006) 'Film tourism and destination marketing: the case of Captain Corelli's Mandolin', *Journal of Vacation Marketing*, 12(3), pp. 256–68.

Hughes, G. (1998) 'Tourism and the semiological realization of space', in Ringer, G. (ed.), *Destinations: Cultural Landscapes in Tourism*, Padstow: Routledge, pp. 17–32.

Huxtable, A. (1997) *The Unreal America: Architecture and Illusion*, New York: New Press.

Ilbery, B. and Bowler, I. (1998) 'From agricultural productivism to post-productivism', in Ilbery, B. (ed.), *The Geography of Rural Change*, Abingdon: Routledge.

Ilbery, B. and Maye, D. (2010) 'Agricultural restructuring and changing food networks in the UK', in Coe, N. and Jones, A. (eds), *The Economic Geography of the UK*, London: Sage, pp. 166–80.

Independent (2016) 'The UK's 10 best theme parks', independent.co.uk, 10 May.

Ioannides, D. (2009) Keynote speech 'The creation of leisure utopias', ATLAS Annual Conference Experiencing difference, Changing Tourism and Tourists' Experiences, Aalborg, Denmark, 27–29 May.

Jackson, E. and Johnson, D. (1991) 'Geographic implications of mega-malls, with specific reference to West Edmonton Mall', *Canadian Geographer*, 35, pp. 226–31.

Jacobs, J. (1961 [2000]) *The Death and Life of Great American Cities*, London: Pimlico.

Jamal, T. and Hill, S. (2002) 'The home and the world: (post)touristic spaces of (in)authenticity?,' in Dann, G. (ed.), *The Tourist as a Metaphor of the Social World*, Oxford: CABI, pp. 77–107.

Jamal, T. and Kim, H. (2005) 'Bridging the interdisciplinary divide: towards an integrated framework for heritage tourism research', *Tourist Studies*, 5(1), pp. 55–83.

Jameson, F. (1985) 'Postmodernism and consumer society', in Foster, H. (ed.), *Postmodern Culture*, London: Pluto, pp. 111–25.

Jameson F. (1998) *The Cultural Turn: Selected Writings on the Postmodern 1983–1998*, London: Verso.

Jansen-Verbeke, M.C. (1990) 'Leisure and shopping: tourism product mix', in Ashworth, G. and Goodall, B. (eds), *Marketing Tourism Places*, London: Routledge, pp. 128–35.

Jansen-Verbeke, M.C. (1991) 'Leisure shopping: a magic concept for the tourism industry', *Tourism Management*, 12, pp. 9–14.

Jansen-Verbeke, M.C. (1999) 'Industrial heritage: a nexus for sustainable tourism development', *Tourism Geographies*, 1(1), pp. 70–85.

Jansson, A. (2010) 'The city in-between: communication geographies, tourism and the urban unconscious', in Knudsen, B. and Waade, A. (2010) *Re-Investing Authenticity Tourism, Place and Emotions*, Bristol: Channel View, pp. 38–51.

Jarratt, D. (2015) 'Seasideness: sense of place at a seaside resort', in Gammon, S. and Elkington, S. (eds), *Landscapes of Leisure: Space, Place and Identities*, Basingstoke: Palgrave, pp.147–63.

Jarratt, D. and Gammon, S. (2016) '"We had the most wonderful times": seaside nostalgia at a British resort', *Tourism Recreation Research*, 41(2), pp. 123–33.

Jeans, D.N. (1990) 'Beach resort morphology in England and Australia: a review and extension', in Fabbri, P. (ed.), *Recreational Uses of Coastal Areas*, Dordrecht: Kluwer.

Jenkins, O. (2003) 'Photography and travel brochures: the circle of representation', *Tourism Geographies*, 5, pp. 305–28.

Jinto (n.d.) Travel blog, http://us.jnto.go.jp/popculture/anime.php?a=1, accessed 18 August 2016.

Johansson, M. and Kociatkiewicz, J. (2011) 'City festivals: creativity and control in staged urban experiences', *European Urban and Regional Studies*, pp. 1–14.

Jokilehto, J. (1999) *A History of Architectural Conservation*, Oxford: Butterworth-Heinemann.

Jokinen, E. and Veijola, S. (1997) 'The disoriented tourist: the figuration of the tourist in contemporary cultural critique', in Rojek, C. and Urry, J. (eds), *Touring Cultures: Transformations of Travel and Theory*, London: Routledge, pp. 23–51.

Jones, T. (2016) 'The future for UK arts, culture and heritage: implementing the new strategy', Westminster Media Forum, 7 June.

Jones, O. and Kirby, S. (2007) *Stately Attraction Film Tourism: Does Genre Influence Interaction?*, London: BFI.

Jordan, J. (2006) *Structures of Memory: Understanding Urban Change in Berlin and Beyond*, Stanford, CA: Stanford University Press.

Josiam, B., Mattson, M. and Sullivan, P. (2004) 'The historaunt: heritage tourism at Mickey's dining car', *Tourism Management*, 25, pp. 453–61.

Judd, D. (1999) 'Constructing the tourist bubble', in Judd, D. and Fairstein, S. (eds), *The Tourist City*, London: Yale University Press, pp. 35–53.

Kaifeng Travel Guide (2017) www.chinahighlights.com/kaifeng/attraction/millennium-city-park.htm, accessed 26 April 2017.

Keith, W.J. (1989) 'The land in Victorian literature', in Mingay, G.E. (ed.), *The Rural Idyll*, London: Routledge.

Kelly, I. (2008) 'Precincts within the urban form: relationships with the city', in Hayllar, B., Griffin, T. and Edwards, D. (eds), *City Spaces – Tourist Places: Urban Tourism Precincts*, Oxford: Butterworth Heinemann.

Kennell, J. (2011) 'Rediscovering cultural tourism: cultural regeneration in seaside towns', *Journal of Town and City Management*, 1, pp. 364–80.

Kim, H. and Jamal, T. (2007) 'Touristic quest for existential authenticity', *Annals of Tourism Research*, 34(1), pp.181–201.

Kirshenblatt-Gimblett, B. (1998) *Destination Culture: Tourism, Museums, and Heritage*, Berkeley: University of California Press.

Klingmann, A. (2007) *Brandscapes: Architecture in the Experience Economy*, Cambridge, MA: MIT Press.

Kolar, T. and Zabkar, V. (2010) 'A consumer based model of authenticity: an oxymoron or the foundation of cultural heritage?', *Tourism Management*, 31(5), pp. 652–64.

Kotler, P. (1986) *Principles of Marketing*, Englewood Cliffs, NJ: Prentice Hall.

Knudsen, B. and Waade, A. (2010) 'Performative authenticity in tourism and spatial experience: rethinking the relations between travel, place and emotion', in Knudsen, B. and Waade, A. (eds), *Re-Investing Authenticity Tourism, Place and Emotions*, Bristol: Channel View, pp. 1–21.

Krolikowski, C. and Brown, G. (2010) 'The structure and form of urban tourism precincts: setting the stage for tourist performances', in Hayllar, B., Griffin, T. and Edwards, D. (eds), *City Spaces-Tourist Places*, London: Routledge, pp. 127–45.

Kürti, L. (2004) 'Picture perfect: community and commemoration in postcards', in Alfonso, A.I., Kurti, L. and Pink, S. (eds), *Working Images: Visual Research and Representation in Ethnography*, London: Routledge, pp. 47–71.

Kwon, M. (2004) *One Place after Another: Site-Specific Art and Locational Identity*, Cambridge, MA: MIT Press.

Labadi, S. (2010) 'World Heritage authenticity and post authenticity: national and international perspectives', in Labadi, S. and Long, C., *Heritage and Globalisation*, Abingdon: Routledge, pp. 66–84.

Laderman, D. (1996) 'What a trip: the road film and American culture', *Journal of Film and Video*, pp. 41–57.

Lakoff, G. and Johnson, M. (2008) *Metaphors We Live By*, Chicago: University of Chicago Press.

Landry, C. (2000) *The Creative City: A Toolkit for Urban Innovators*, London: Earthscan.

Langen S. (2015) *Outdoor Art: Extraordinary Sculpture Parks and Art in Nature*, New York: Prestel.

Larsen, J. (2005) 'Families seen sightseeing: performativity of tourist photography', *Space and Culture*, 8, pp. 416–34.

Larsen, J. (2008) 'Practices and flows of digital photography: an ethnographic framework', *Mobilities*, 3(1), pp. 141–60.

Larsen, J. and Urry, J. (2011) 'Gazing and performing', *Environment and Planning D: Society and Space*, 29(6), pp. 1110–25.

Las Vegas Convention and Visitors Authority (2016) *Stats and Facts: Population of Las Vegas*, lvcva.com, accessed 24 April 2017.

Law, C.M. (1993) *Urban Tourism: Attracting Visitors to Large Cities*, London: Mansell.

Law C.M. (ed.) (1996) *Tourism in Major Cities*, London: International Thompson Business.

Lefebvre, H. (1974, 1991, 2006) *The Production of Space*, Oxford: Blackwell.

Lefebvre, M. (2006) *Landscape and Film*. London: Taylor & Francis.

Lehtonen, T.-K. and Mäenpää, P. (1997) 'Shopping in the East Centre Mall', in Falk, P. and Campbell, C. (eds), *The Shopping Experience*, London: Sage.

Lencek, L. and Bosker, G. (1999) *The Beach: The History of Paradise on Earth*, London: Penguin.

Lew, A., Hall, C.M. and Timothy D.J. (2008) *World Geography of Travel and Tourism: A Regional Approach*, Oxford: Butterworth-Heinemann.

Ley, D. (1996) *The New Middle Class and the Remaking of the Central City*, New York: Oxford University Press.

Lippard, L. (1997) *The Lure of the Local: Senses of Place in a Multicentred Society*, New York: New Press.

Littlejohn, D. (ed) (1999) *The Real Las Vegas: Life Beyond the Strip*, New York: Oxford University Press.

Littrell, M., Anderson, L. and Brown, P. (1993) 'What makes a craft souvenir authentic?', *Annals of Tourism Research*, 20, pp. 197–215.

Lloyd, R. (2010) *Neo Bohemia: Art and Commerce in the Postindustrial City*, Abingdon: Routledge.

Locke, A. (2015) 'Catalogue plates', *Ai Wei Wei*, London: Royal Academy of Arts.

Lockhart, D. and Drakakis-Smith, D.W. (eds) (1997) *Island Tourism: Trends and Prospects*, London: Cassell.

Loewen, J. (1999) *Lies Across America: What Our Historic Sites Get Wrong*, New York: New Press.

London Resort Company Holdings (n.d.) http://www.londonparamount.info/about-us/, accessed 19 December 2016, 09:00.

Lovell, J. (2003) *Canterbury Visitor Survey*, Canterbury: Canterbury City Council.

Lovell, J. (2005) Cathédrales de Lumière, Interreg project with Canterbury, Rochester, Amiens and Rouen cathedrals.

Lovell, J. (2009) 'Post-modern heritage, chivalry, park and ride: Le Tour comes to Canterbury', in Ali-Knight, J. (ed.), *International Perspectives of Festivals and Events: Paradigms of Analysis*, London: Routledge, pp. 265–78.

Lovell, J. (2010) 'Historic cities: from real to fake', in Lovell, J. and Stuart-Hoyle, M. (eds), *Leisure Experiences: Space, Place and Performance*, Leisure Studies Association, 109.

Lovell, J. (2013) 'Cities of imagination: staged authenticity in historic cities', doctoral dissertation, University of Kent.

Lovell, J. and Beedie, C. (2011) 'Placebo heritage', draft of a research note.

Lovell, J. and Brown, A. (2014) Transformation Agenda Working Group, Kent Heritage and Culture.

Lovell, J. and Hitchmough, S. (2015) 'An American yell: Buffalo Bill and the Wild West in East Kent', Research Without Borders Symposium, Canterbury Christ Church University, July.

Lowe, P. (1989) 'The rural idyll defended: from preservation to conservation', in Mingay, G.E. (ed.), *The Rural Idyll*, London: Routledge.

Lowe, P.D. and Goyder, J.M. (1983) *Environmental Groups in Politics*, London: Allen and Unwin.

Lowenthal, D. (1985) *The Past is a Foreign Country*, Cambridge: Cambridge University Press.

Lowenthal, D. (1998) *The Heritage Crusade and the Spoils of History*, Cambridge: Cambridge University Press.

Lowerson, J. and Myerscough, J. (1977) *Time to Spare in Victorian England*, Hassocks: Harvester Press.

Lycett Green, C. (2011) *Seaside Resorts*, London: Oldie.

Lynch, K. (1972) *What Time is this Place?*, Cambridge, MA: MIT Press.

Lynch, D. (2014) http://www.urbansplash.co.uk/blog/urban-splash/english-heritage-use-park-hill-as-an-exemplar-of-regeneration/, accessed 7 January 2017, 10:00.

Mabey, R (2000) *Country Matters: Selected Writings 1974–1999*, London: Pimlico.

MacCannell, D. (1973 [1999]) 'Staged authenticity: arrangements of social space in tourist settings', *American Sociological Review*, 79, pp. 589–603.

MacCannell, D. (1976) *The Tourist: A New Theory of the Leisure Class*, New York: Solouker.

MacCannell, D. (1989) *The Touris:. A New Theory of the Leisure Class*, New York: Schocken.

MacDonald, G. and Alsford, S. (1995) 'Museums and theme parks: worlds in collision?', *Museum Management and Curatorship*, 14(2), pp. 129–47.

MacDonald, S. (2009) *Difficult Heritage: Negotiating the Nazi Past in Nuremberg and Beyond*, Abingdon: Routledge.

Macfarlane, R. (2003) *Mountains of the Mind*, London: Granta.

Macfarlane, R. (2008) *The Wild Places*, London: Granta.

Macfarlane, R. (2015) *Landmarks*, London: Hamish Hamilton.

MacLeod, N.E. (2006) 'The placeless festival: identity and place in the post-modern festival', in Picard, D. and Robinson, M. (eds), *Festivals, Tourism and Social Change: Remaking Worlds*, Vol. 8, Bristol: Channel View, pp. 222–37.

Maitland, R. (2007) 'Tourists, the creative class and distinctive areas in major cities: the role of visitors and residents in developing new tourism areas', in Richards, G. and Wilson, J. (eds), *Tourism Creativity and Development*, Abingdon: Routledge, pp. 73–86.

Maitland, R. (2010) 'Everyday life as a creative experience in cities', *International Journal of Culture, Tourism and Hospitality Research*, 4(3), pp. 176–85.
Maitland, R. and Newman, P. (2014) *World Tourism Cities: Developing Tourism Off the Beaten Track*, London: Routledge.
Maitland, R. and Smith, A. (2009) 'Tourism and the aesthetics of the built environment', in Tribe, J. (ed.), *Philosophical Issues in Tourism (Aspects of Tourism)*, 37, Bristol: Channel View, pp. 171–90.
Manning, F.E. (1992) 'Spectacle', in Bauman, R. (ed.), *Folklore, Cultural Performances and Popular Entertainments: A Communications-Centred Handbook*, Oxford: Oxford University Press, pp. 291–9.
Maoz, D. (2006) 'The mutual gaze', *Annals of Tourism Research*, 33(1), pp. 221–39.
Massey, D. (2005) *For Space*, London: Sage.
Massey, D. (2013) *Space, Place and Gender*, New Jersey: John Wiley & Sons.
Massey, D., Allen, J. and Sarre, P. (eds) (1999) *Human Geography Today*, Oxford: Blackwell.
Matless, D. (1990) 'Definitions of England, 1928–89: preservation, modernism and the nature of the nation', *Built Environment*, 16(3).
Matthews, H., Taylor, M., Sherwood, K., Tucker, F. and Limb, M. (2000) 'Growing-up in the countryside: children and the rural idyll', *Journal of Rural Studies*, 16, pp. 141–53.
Mathews, V. (2010) 'Aestheticizing space: art, gentrification and the city', *Geography Compass*, 4(6), pp. 660–75.
Mathews, V. (2014) 'Incoherence and tension in culture-led redevelopment', *International Journal of Urban and Regional Research*, 38(3), pp. 1019–36.
McCarthy, J. (2008) 'More than just telling the story: interpretive narrative archaeology', in Fairclough, G., Harrison, R., Jameson, J. and Schofield, J. (eds), *The Heritage Reader*, Abingdon: Routledge.
McEwen, A. and McEwen, M. (1982) *National Parks, Conservation or Cosmetics?*, London: Allen and Unwin.
McLean, K. (2016) 'Smellmap: Amsterdam – olfactory art and smell visualisation', *Leonardo*, 50(1), pp. 92–3.
Meethan, K. (1996) 'Consuming (in) the civilised city', *Annals of Tourism Research*, 23(2), pp. 322–40.
Merrifield, A. (2000) 'Henri Lefebvre: a socialist in space', in Crang, M. and Thrift, N. (eds), *Thinking Space*, London: Routledge, pp. 167–82.
Message, K. (2006) *New Museums and the Making of Culture*, Oxford: Berg.
Mingay, G.E. (ed.) (1989) *The Rural Idyll*, London: Routledge.
Monbiot, G. (2013) 'A manifesto for rewilding the world', *Guardian*, 28 May.
Monbiot, G. (2014) *Feral: Rewilding the Land, Sea and Human Life*, Chicago: University of Chicago Press.
Monro, R. (1996) 'The consumption view of self: extension, exchange and identity', in Edgell, S., Hetherington, K. and Warde, A. (eds), *Consumption Matters*, Oxford: Blackwell, pp. 248–73.
Mooney, G. and Danson, M. (1997) 'Beyond "culture city": Glasgow as a dual city', in Jewson, N. and MacGegor, S. (eds), *Transforming Cities: Contested Governance and New Spatial Divisions*, London: Routledge, pp. 73–86.
Morgan, N.J. and Pritchard, A. (1999) *Power and Politics at the Seaside*, Exeter Maritime Studies, Exeter: University of Exeter Press.
Moscardo, G. and Pearce, P. (1986) 'Historic theme parks: an Australian experience in authenticity', *Annals of Tourism Research*, 13(3), pp. 467–79.

Moseley, M.J. (1979) *Accessibility: The Rural Challenge*, London: Methuen.

Moseley, P. (2012) 'Royale de Luxe, a phenomenon in street theatre', in Liverpool City Council, *In The Footsteps of Giants*, Liverpool: Liverpool University Press.

Mount, H. (2013) *How England Made the English*, London: Penguin.

Mowforth, M. and Munt, I. (1998) *Tourism and Sustainability: New Tourism in the Third World*, London: Routledge.

Muggleton, D. and Weinzierl, R. (2003) *The Post-Subcultures Reader*, Oxford: Berg.

Mules, T. (1998) 'Events tourism and economic development in Australia', in Tyler, D., Guerrier, Y. and Robertson, M. (eds), *Managing Tourism in Cities: Policy, Process and Practice*, Chichester: John Wiley & Sons.

Murfitt, N. (2015) 'It's the great Dad's Army home front makeover: the theatres, butchers and pubs of East Yorkshire have been transformed into the fictional town of Walmington-on Sea for new movie', *Daily Mail*, http://www.dailymail.co.uk/news/article-3157816/It-s-great-Dad-s-Army-home-makeover-theatres-butchers-pubs-East-Yorkshire-transformed-fictional-town-Walmington-Sea-new-movie.html, accessed 10 August 2015, 16:25.

Myerscough, J. (1974) 'The recent history of the use of leisure time', in Appleton, I. (ed.), *Leisure Research and Policy*, Edinburgh: Scottish Academic Press.

Nash, C. (2000) 'Performativity in practice: some recent work in cultural geography', *Progress in Human Geography*, 24, pp. 653–64.

Nash, P. (1967) *Wilderness and the American Mind*, London: Yale University Press.

National Trust (2014) *National Trust: Annual Report 2013/14*, London: National Trust.

Nettleton, J. and Shapiro, S. (2016) 'The future for UK arts, culture and heritage: implementing the new strategy', Westminster Media Forum, 7 June.

Newby, H. (1980) *Green and Pleasant Land*, Harmondsworth: Pelican.

Newby, H. (1988) *The Countryside in Question*, London: Hutchinson.

Obrador Pons, P. (2003) 'Being-on-holiday: tourist dwelling, bodies and place', *Tourist Studies*, 3, pp. 47–66.

Olick, J. (2003) *States of Memory: Continuities, Conflicts and Transformations in National Retrospection*, Durham, NC: Duke University Press.

Orbaşli, A. (2000) *Tourists in Historic Towns: Urban Conservation and Heritage Management*, London: Taylor & Francis.

Otero-Pailos, J., West, S. and Gaiger, J. (2010) 'Heritage values', in West, S. (ed.), *Understanding Heritage in Practice*, Manchester: Manchester University Press, pp. 47–87.

Padilla-Meléndez, A. and del Águila-Obra, A.R. (2013) 'Web and social media usage by museums: online value creation', *International Journal of Information Management*, 33(5), pp. 892–98.

Page, S. (1995) *Urban Tourism*, London: Routledge.

Page, S. and Hall, C. (2003) *Managing Urban Tourism*, Harlow: Prentice-Hall.

Page, S.J. and Connell, J. (2010) *Leisure: An Introduction*, Harlow: Pearson Education.

Pahl, R.E. (1965) *Urbs in Rure: The Metropolitan Fringe in Hertfordshire*, Paper No. 2 London School of Economics and Political Science.

Pallagst, K., Wiechmann, T. and Martinez-Fernandez, C. (2013) *Shrinking Cities: International Perspectives and Policy Implications*, London: Routledge.

Pallant C. (2013) 'The animated landscape', in Rayner, J. and Harper, G. (eds), *Film Landscapes: Cinema, Environment and Visual Culture*, Newcastle: Cambridge Scholars.

Palmer-RAE Associates (2004) *European Cities and Capitals of Culture*, Brussels: Palmer-RAE Associates.

Parr, M. (n.d.) Introduction, https://www.martinparr.com/introduction/, accessed 23 April 2017, 20:00.

Patmore, J.A. (1968) 'The spa towns of Britain', in Beckinsale, R.P. and Houston, J.M. (eds), *Urbanisation and its Problems*, Oxford: Blackwell.

Paxman, J. (1998) *The English: Portrait of a People*, London: Michael Joseph.

Pearce, D. (1995) *Tourism Today: A Geographical Analysis*, 2nd edn, Harlow: Longman.

Pernecky, T. and Jamal, T. (2010) '(Hermeneutic) phenomenology in tourism studies', *Annals of Tourism Research*, 37(4), pp.1055–75.

Perry, G., Lowe, A. and Moore, S. (2013) *The Vanity of Small Differences*, London: Hayward.

Perry, J. and Croft, D. (n.d.) http://www.walmington-on-line.co.uk/home/visit-walmington-on-sea/, accessed 10 August 2016, 17:05.

Pheasant-Kelly, F.E. (2014) 'Landscape, spectacle and subjectivity: abject spaces in Peter Jackson's Lord of the Rings Trilogy', in Rayner, J. and Harper, G. (eds), *Film Landscapes: Cinema, Environment and Visual Culture*, Cambridge: Cambridge Scholars.

Phillips, M. and Smith, D. (2015) 'Rural gentrification and the class complexion of the British countryside 10 years on', paper delivered at the Royal Geographical Society (with IBG), Annual International Conference, University of Exeter.

Picard, D. and Robinson, M. (eds) (2006) *Festivals, Tourism and Social Change: Remaking Worlds*, Bristol: Channel View.

Pikkemaat, B. and Schuckert, M. (2007) 'Success factors of theme parks: an exploratory study', *Preliminary Communication*, 55(2), pp. 197–208.

Pile, S. (2005) *Real Cities: Modernity, Space and the Phantasmagorias of City Life*, London: Sage.

Pine, B. and Gilmore, H. (1999) *The Experience Economy: Work is Theatre and Every Business a Stage*, Boston: Harvard University Press.

Pink, J. (2001) *Doing Visual Ethnography*, London: Sage.

Pink, J. (2007) 'Mobilising visual ethnography: making routes, making place and making images', keynote paper at University of Brighton Conference, Gazing, Glancing and Glimpsing?: Tourists and Tourism in a Visual World, 13–15 June.

Pink, S., Hubbard, P., O'Neill, M. and Radley, A. (2010) 'Walking across disciplines: from ethnography to arts practice', *Visual Studies*, 25(1), pp. 1–7.

Plog, S. (1974) 'Why destination areas rise and fall in popularity', *Cornell Hotel and Restaurant Administration Quarterly*, 14(4), pp. 55–8.

Plutarch and Perrin, B. (trans.) (1998) *Plutarch's Lives: Theseus and Romulus, Lycurgus and Numa, Solon and Publicola*, Cambridge, MA: Harvard, Loeb Classical Library.

Poria, Y., Butler, R. and Airey, D. (2003) 'The core of heritage tourism', *Annals of Tourism Research*, 30(1), pp. 238–54.

Pred, A. (1984) 'Place as historically contingent process: structuration and the time-geography of becoming places', *Annals of the Association of American Geographers*, 74(2), pp. 279–97.

Prosser, J. (1998) *Image-based Research: A Sourcebook for Qualitative Researchers*, Abingdon: Psychology Press.

Quinn, B. (2005) 'Arts festivals and the city', *Urban Studies*, 42(5/6), pp. 927–43.

Quinn, B. and Wilks, L. (2016) *Linking Social Capital, Cultural Capital and Heterotopia at the Folk Festival*, Dublin: Dublin Institute of Technology.

Raban, J. (1974) *Soft City*, Glasgow: Fontana.

Rackham, O. (1986) *The History of the Countryside*, London: Phoenix.

Raento, P. and Flusty, S. (2006) 'Three trips to Italy: deconstructing the new Las Vegas', in Minca, C. and Oakes, T. (eds), *Travels in Paradox: Remapping Tourism*, Lanham, MD: Rowman and Littlefield.

Ram, Y, Björk, P. and Weidenfeld, A. (2015) 'Authenticity and place attachment of major visitor attractions', *Tourism Management*, 52, pp. 110–22.

Ramblers' Association (2016) *The Ramblers' Big Pathwatch Survey of Footpaths in England and Wales*, London: Ramblers' Association.

Rayner, J. and Harper, G. (eds) (2014) *Film Landscapes: Cinema, Environment and Visual Culture*, Cambridge: Cambridge Scholars.

Reisinger, Y. and Steiner, C. (2005) 'Understanding existential authenticity', *Annals of Tourism Research*, 33(2), pp. 299–318.

Reisinger, Y. and Steiner, C. (2006) 'Reconceptualising interpretation: the role of tour guides in authentic tourism', *Current Issues in Tourism*, 9(6), pp. 481–98.

Relph, E. (1976 [2008]) *Place and Placelessness*, London: Pion.

Richards, G. (1996) 'Production and consumption of European tourism', *Annals of Tourism Research*, 23(2), pp. 261–83.

Richards, G. (2007) 'Culture and authenticity in a traditional event: the views of producers, residents, and visitors in Barcelona', *Event Management*, 11(1–2), pp. 33–44.

Richards, G. (2010) *Leisure in the Network Society: From Pseudo Events to Hyperfestivity?*, Tilburg: Tilburg University, Researchgate, https://www.researchgate.net/profile/Greg_Richards3/publication/277042387_Festivals_in_the_network_society/links/5560414a08ae6f4dcc927ee1.pdf, accessed 28 November 2016, 12:15.

Richards, G. (2014) 'Imagineering events as interaction ritual chains', in Richards, G., Marques, L. and Mein, K. (eds), *Event Design: Social Perspectives and Practices*, Abingdon: Routledge.

Richards, G. (2015) 'Events in the network society: the role of pulsar and iterative events', *Event Management*, 19(4), pp. 553–66.

Richards, G. and Palmer, R. (2010) *Eventful Cities: Cultural Management and Urban Revitalisation*, Oxford: Butterworth Heinemann.

Richards, G. and Wilson, J. (2006) 'Developing creativity in tourist experiences: a solution to the serial reproduction of culture?', *Tourism Management*, 27(6), pp. 1209–23.

Richards, G. and Wilson, J. (2007) 'The creative turn in regeneration: creative spaces, spectacles and tourism in cities', in Smith, M., *Tourism, Culture and Regeneration*, Wallingford: CABI.

Ritzer, G. and Liska, A. (1997) '"McDisneyization" and "post-tourism": complementary perspectives on contemporary tourism', in Rojek C. and Urry J. (eds), *Touring Cultures: Transformations of Travel and Theory*, London: Routledge, pp. 96–112.

Robertson, M. and Guerrier, Y. (1998) 'Events as entrepreneurial displays: Seville, Barcelona and Madrid', in Tyler, D., Guerrier, Y. and Robertson, M. (eds), *Managing Tourism in Cities: Policy, Process and Practice*, Chichester: Wiley.

Robertson, M., Chambers, D. and Frew, E. (2007) 'Events and festivals: current trends and issues', *Managing Leisure*, 12(2–3), pp. 91–101.

Robertson Scott, J.W. (1925) *England's Green and Pleasant Land*, Harmondsworth: Penguin.

Robinson, H. (1976) *A Geography of Tourism*, London: Macdonald & Evans.

Robinson, M. (2002) 'Between and beyond: literature-tourism relationships', in Robinson, M. and Anderson, H.C. (eds), *Literature and Tourism: Essays in the Reading and Writing of Tourism*, London: Thomson.

Robinson, M. and Anderson, H.C. (2002) 'Literature and the creation of tourist spaces', in Robinson, M. and Anderson, H.C. (eds), *Literature and Tourism: Essays in the Reading and Writing of Tourism*, London: Thomson.
Robinson, M. and Picard, D. (eds) (2009) *The Framed World: Tourism, Tourists and Photography*. Farnham: Ashgate.
Roesch, S. (2009) *The Experiences of Film Location Tourists*, Vol. 42, Bristol: Channel View.
Rojek, C. (1997) 'Indexing, dragging and the social construction of tourist sites', in Rojek, C. and Urry, J. (eds), *Touring Cultures: Transformations of Travel and Theory*, Abingdon: Psychology Press, pp. 52–74.
Roodhouse, S. (2006) *Cultural Qu4rters: Principles and Practices.* Bristol: Intellect.
Rosbottom, D. (2015) 'Architecture can also be silent', in Royal Academy of Arts, *Ai Wei Wei*, London: Royal Academy of Arts.
Roscoe, B. (1996) '"Bradford-on-Avon but Shell on the road": the heyday of motor touring through Britain's countryside', in Watkins, C. (ed.), *Rights of Way: Policy, Culture and Management*, London: Pinter.
Rose, G. (2001) *Visual Methodologies: An Introduction to the Interpretation of Visual Methodologies*, London: Sage.
Rose, G. (2007) 'Researching visual materials: towards a critical visual methodology', reprinted edn, *Visual Methodologies. An Introduction to Interpreting Visual Materials*. London: Sage, pp. 1–32.
Rothman, H.K. and Davis, M. (eds) (2002) *The Grit beneath the Glitter: Tales from the Real Las Vegas*, Berkeley: University of California Press.
Rowley, T. (1994) *Villages in the Landscape*, London: Orion.
Rowling, J.K. (1999) *Harry Potter and the Prisoner of Azkaban*, London: Bloomsbury.
Royale de Luxe (n.d.) http://www.giantspectacular.com, accessed 4 December 2016, 17:10.
RSPB (2016) *What We Do: Timeline and Milestones*, www.rspb.org.uk, accessed 24 April 2017.
Runte, A. (1997) *National Parks: The American Experience*, Lincoln: University of Nebraska Press.
Russo, A. (2002) 'The "vicious circle" of tourism development in heritage cities', *Annals of Tourism Research*, 29(1), pp. 165–82.
Russell, H. (2013) *Viking Heritage to Get Rebrand with Disney-Style Amusement Park*, https://www.theguardian.com/world/2013/dec/06/viking-heritage-amusement-park-western-norway, accessed 19 December 2016, 09:15.
Ryan, C. (1991) *Recreational Tourism: A Social Science Perspective*, London: Routledge.
Ryan, C. (ed.) (1997) *The Tourist Experience: A New Introduction*, London: Cassell.
Ryan, C. (2002) 'Memories of the beach', in Ryan, C. (ed.), *The Tourist Experience: A New Introduction*, 2nd edn, London: Continuum.
Salamone, F. (1997) 'Authenticity in tourism: the San Angel', *Annals of Tourism Research*, 24(2), pp. 305–21.
Salazar, N. (2010) 'The globalisation of heritage through tourism: balancing standardisation and differentiation', in Labadi, S. and Long, C. (2010) *Heritage and Globalisation*, Abingdon: Routledge, pp. 130–46.
Samuel, R. (1994) *Theatres of Memory*, Vol. 1, London: Verso.
Sanneh, K. (2016) 'Is gentrification really a problem?', *New Yorker*, http://www.newyorker.com/magazine/2016/07/11/is-gentrification-really-a-problem, accessed 23 October 2016, 16:00.

Schwenger, P. (2006) *The Tears of Things: Melancholy and Physical Objects*, London: University of Minnesota Press.

Seamon, D. (1980) 'Body-subject, time-space routines and place ballet', in Buttimer, A. and Seamon, D., *The Human Experience of Space and Place*, London: Croom Helm.

Selby, M. (2004) *Understanding Urban Tourism: Image, Culture and Experience*, London: I.B. Tauris.

Selwyn, T. (1996) 'Introduction', *The Tourist Image: Myths and Myth Making in Tourism*, Chichester: Wiley.

Shackley, M. (2001) *Managing Sacred Sites*, New York: Continuum.

Sharpley, R. and Sharpley, J. (1997, 2007) *Tourism and Leisure in the Countryside*, London: International Thompson Business.

Shaw, G. and Williams, A.M. (eds) (1997) *The Rise and Fall of British Coastal Resorts*, London: Pinter.

Shaw, G. and Williams, A.M. (2002) *Critical Issues in Tourism: A Geographical Perspective*, 2nd Edition. Oxford: Blackwell.

Shaw, G. and Williams, A.M. (2004) *Tourism and Tourism Spaces*, London: Sage.

Shearing, C. and Stenning, P. (2002) 'From the panoptican to Disney World: the development of discipline', in McLaughlin, E., Muncie, J. and Hughes, G. (eds), *Criminal Perspectives: Essential Readings*, London: Sage.

Shields, R. (1990) '"The system of pleasure": liminality and the carnivalesque at Brighton', *Theory, Culture and Society*, 7(1), pp. 39–72.

Shields, R. (1990) *Places on the Margin: Alternative Geographies of Modernity*, London: Routledge.

Shoard, M. (1980) *The Theft of the Countryside*, London: Maurice Temple Smith.

Shoard, M. (1982) 'The lure of the moors', in Gold, J.R. and Burgess, J. (eds), *Valued Environments*, London: George Allen & Unwin.

Shoard, M. (1987) *This Land is Our Land*, London: Paladin.

Shoard, M. (1999) *A Right to Roam: Should We Increase Access to Britain's Countryside*, Oxford: Oxford University Press.

Shoard, M. (2000) 'Edgelands of promise', *Landscapes*, 1(2), pp. 74–93.

Shoard, M. (2002) 'Edgelands', in Jenkins, J. (ed.), *Remaking the Landscape*, London: Profile.

Sibley, D. (1988) 'Survey 13: purification of space', *Environment and Planning D: Society and Space*, 6, pp. 409–21.

Siggins, L. (2016) 'Concern over "incidents" during "Star Wars" filming on Skellig Michael', *The Irish Times*, http://www.irishtimes.com/culture/film/concern-over-incidents-during-star-wars-filming-on-skellig-michael-1.2532753, accessed 14 August 2016, 16:51.

Silvanto, S. and Hellman, T. (2005) 'Helsinki: the festival city', in Lankinen, L. (ed.), *Arts and Culture in Helsinki*, Helsinki: City of Helsinki, pp. 4–9.

Simons, I. (2014) 'How to slay a dragon slowly: applying slow principles to event design', in Richards, G., Marques, L. and Mein, K. (eds), *Event Design: Social Perspectives and Practices*, Abingdon: Routledge.

Smith, C. (1989) 'Museums, artefacts and meanings', in Vergo, P. (ed.), *The New Museology*, London: Reaktion, pp. 6–21.

Smith, L. (2006) *Uses of Heritage*, Abingdon: Routledge.

Smith, M. (2003) *Issues in Cultural Tourism Studies*, London: Routledge.

Smith, M.K. (2004) 'Seeing a new side to seasides: culturally regenerating the English seaside town', *International Journal of Tourism Research*, 6, pp. 17–28.

Smith, N. (1996) *The New Urban Frontier: Gentrification and the Revanchist City*, Abingdon: Psychology Press.

Smith, R. and Bugni, V. (2002) 'Symbolic interaction theory and architecture', *Symbolic Interaction*, 29(2), pp. 123–55.

Soja, E. (1996) *Thirdspace: Journeys to Los Angeles and Other Real-and-Imagined Places*, Oxford: Blackwell.

Soja, E. (1999) 'Thirdspace: expanding the scope of the geographical imagination', in Massey, D., Allen, J. and Sarre, P. (eds), *Human Geography Today*, Oxford: Blackwell.

Soja, E.W. (2010) 'Six discourses on the postmetropolis', in Bridge, G. and Watson, S. (eds), *The Blackwell City Reader*, New Jersey: John Wiley & Sons, pp. 374–82.

Sontag, S. (1977) *On Photography*, London: Penguin.

Sorkin, M. (ed.) (1992) 'Introduction: variations on a theme park', in Sorkin, R., *Variations on a Theme Park: The New American City and the End of Public Space*, New York: Hill and Wang.

Standeven, J. and De Knop, P. (1999) *Sport Tourism*, Champaign, IL: Human Kinetics.

Steiner, C. (1999) 'Authenticity, repetition, and the aesthetics of seriality', in Phillips, R. and Steiner, C. (eds), *Unpacking Culture: Art and Commodity in Colonial and Postcolonial Worlds*, Berkeley: University of California Press, pp. 87–103.

Stevenson, D. (2003) *Cities and Urban Culture*, Maidenhead: Open University Press.

Stone, C. (2009) 'The British pop music festival phenomenon', in Ali-Knight, J., Robertson, M., Fyall, A. and Ladkin, A. (eds), *International Perspectives of Festivals and Events*, Oxford: Elsevier.

Strong, R. (2012) *Visions of England*, London: Vintage.

Stuart-Hoyle, M. and Lovell, J. (2007), 'Liberating the heritage city: towards cultural engagement', in Robinson, M. and Smith, M. (eds), *Cultural Tourism*, Bristol: Channel View, pp. 290–303.

Swarbrooke, J. (2000) 'Museums: theme parks of the third millennium', *Tourism and Heritage Relationships*, pp. 417–31.

Taussig, M.T. (1993) *Mimesis and Alterity: A Particular History of the Senses*, Abingdon: Psychology Press.

TEA/AECOM (2015) *2014 Theme Index and Museum Index: Global Attractions Attendance Report*, Themed Entertainment Association, teaconnect.org, accessed 24 April 2017.

Telegraph (2014) 'The race to stop Las Vegas from running dry', 28 June.

Telegraph (2016) '14 Venices around the world: can you name them?', 25 January.

The Canadian Encyclopedia (2015) 'West Edmonton Mall', thecanadianencyclopedia.ca, 8 July.

Thompson, F. (1939) *Lark Rise to Candleford*, Oxford: Oxford University Press.

Thrift, N. (1989) 'Images of social change', in Hamnett, C., McDowell, L. and Sarre, P. (eds), *The Changing Social Structure*, London: Sage, pp. 12–42.

Thrift, N. (2004) 'Performance and performativity: a geography of unknown lands', in Duncan, J.S., Johnson, N.C. and Schein, R. (eds), *A Companion to Cultural Geography*, Oxford: Blackwell, pp. 121–36.

Thurley, S. (2010) *Romancing the Stone*, BBC, English Heritage, Park Hill, 1 May.

Timothy, D.J. (2001) *Tourism and Political Boundaries*, London: Routledge.

Timothy, D. (2005) *Shopping Tourism, Retailing and Leisure*, Bristol: Channel View.

Towner, J. (1996) *An Historical Geography of Recreation and Tourism in the Western World 1540–1940*, Chichester: Wiley.

transparencynow.com (2016) *Travelling Through Hyperreality With Umberto Eco*, accessed August 2016.

Treble, R. (1989) 'The Victorian picture of the country', in Mingay, G.E. (ed.), *The Rural Idyll*, London: Routledge.

Trilling, L. (1972) *Sincerity and Authenticity*, London: Oxford University Press.

Tuan, Yi-Fu (1974) *Topophilia: A Study of Environmental Perception, Attitudes, and Values*, Englewood Cliffs: Prentice-Hall.

Tuan, Y.F. (1977) *Space and Place: The Perspective of Experience*, Minneapolis: University of Minnesota Press.

Tuan, Y. (2001) *Space and Place*, reprint edn, Minneapolis: University of Minnesota Press.

Tully, M. (2010) 'Alone on the moors', in Bryson, B. (ed.), *Icons of England*, London: Transworld.

Tunbridge, J. and Ashworth, G. (1996) *Dissonant Heritage: The Management of the Past as a Resource in Conflict*, Chichester: Wiley.

Tunstall, S.M. and Penning-Rowsell, E.C. (1998) 'The English beach: experiences and values', *The Geographical Journal*, 164(3), pp. 319–32.

Turner, C. and Manning, P. (1988) 'Placing authenticity-on being a tourist: a reply to Pearce and Moscardo', *Australia and New Zealand Journal of Sociology*, 24, pp. 136–9.

Turner, J. and Hodgson, J. (1979) 'Studies in the vegetational history of the northern Pennines: I, Variations in the composition of the Early Flandrian forests', *Journal of Ecology*, 67(2), pp. 629–46, cited in Rackham, O. (1986) *The History of the Countryside*, London: Phoenix.

Turner, V. and Turner, E. (1978) *Image and Pilgrimage in Christian Culture*, New York: Columbia University Press.

UNESCO (1994) Nara Document on Authenticity, http://whc.unesco.org/archive/opguide08-en.pdf, accessed 28 September 2010, 17:00.

UNESCO (2007) [Online] Nara Document on Authenticity, http://whc.unesco.org/uploads/events/documents/event-833-3.pdf, accessed 1 January 2013.

UNESCO (2011) http://whc.unesco.org/archive/opguide08-en.pdf, p. 32, accessed 28 September 2010, 17:00.

UNESCO Operational Guidelines for the Implementation of the World Heritage Convention (2008) http://whc.unesco.org/archive/opguide08-en.pdf, accessed 10 April 2017, 18:00.

Universal Studios (n.d.) *A Look at the Wizarding World of Harry Potter with Supervising Art Director Alan Gilmore Inside Universal*, http://insideuniversal.net/2016/03/a-look-at-the-wizarding-world-of-harry-potter-with-supervising-art-director-alan-gilmore/, accessed 30 December 2016, 09:30.

Urry, J. (1995a) *Consuming Places*, London: Routledge

Urry, J. (1995b) 'How societies remember the past', *The Sociological Review*, 43(S1), pp. 45–65.

Urry, J. (1995c) 'A middle class countryside', in Butler, T. and Savage, M. (eds), *Social Change and the Middle Classes*, London: UCL Press.

Urry, J. (1997) 'Cultural change and the seaside resort', in Shaw, G. and Williams, A.M. (eds), *The Rise and Fall of British Coastal Resorts*, London: Pinter.

Urry, J. (2002) *The Tourist Gaze*, 2nd edn, London: Sage.

Urry, J. and Larsen, J. (2011) *The Tourist Gaze 3.0*, London: Sage.

Uzzell, D. and Ballantyne, R. (1998) *Contemporary Issues in Heritage Interpretation: Problems and Prospects*, London: Stationery Office.

Uzzell, D. and Ballantyne, R. (2008) 'Heritage that hurts: interpretation in a postmodern world', in Fairclough, G., Harrison, R., Jameson, J. and Schofield, J. (eds), *The Heritage Reader*, Abingdon: Routledge.

Valentine, G. (1997) 'A safe place to grow up? Parenting, perceptions of children's safety and the rural idyll', *Journal of Rural Studies*, 13(2), pp. 137–48.

van der Duim, R. (2007) 'Tourismscapes: an actor-network perspective', *Annals of Tourism Research*, 34(4), pp. 961–76.

Venturi, R., Brown, D.S. and Izenour, S. (1972) *Learning from Las Vegas*, Cambridge, MA: MIT Press.

Vogler, C. (1999) *The Writer's Journey*, London: Pan.

Wainwright, O. (2013) 'Seeing double: what China's copycat culture means for architecture', *Guardian*, https://www.theguardian.com/artanddesign/architecture-design-blog/2013/jan/07/china-copycat-architecture-seeing-double, accessed 19 January 2017, 07:35.

Walsh, J. (1992) *The Representation of the Past*, London: Routledge.

Walton, J.K. (1983) *The English Seaside Resort: A Social History1950–1914*, Leicester: Leicester University Press.

Walton, J. (2000) *The British Seaside: Holidays and Resorts in the Twentieth Century*, Manchester: Manchester University Press.

Walton, J. (2013) 'British tourists and the beaches of Europe, from the eighteenth century to the 1960s', in Farr, M. and Guégan, X. (eds), *The British Abroad Since the Eighteenth Century: Vol. 1, Travellers and Tourists*, Basingstoke: Palgrave Macmillan.

Walton, J and Browne, P. (eds) (2010) *Coastal Regeneration in English Resorts – 2010*, Lincoln: Coastal Communities Alliance.

Walvin, J. (1978) *Beside the Seaside: A Social History of the Popular Seaside Holiday*, Harmondsworth: Penguin, Chapters 5–7.

Wang, N. (1999) 'Rethinking authenticity in tourism experience', *Annals of Tourism Research*, 26(2), pp. 349–70.

Wang, N. (2000) *Tourism and Modernity: A Sociological Analysis*, Oxford: Pergamon.

Ward, J. (2016) 'Down by the sea: visual arts, artists and coastal regeneration', *International Journal of Cultural Policy*, pp. 1–18.

Ward, M. and Hardy, D. (1986) *Goodnight Campers! The History of the British Holiday Camp*, London: Mansell.

Watkins, C. (ed.) (1996) *Rights of Way: Policy, Culture and Management*, London: Pinter.

Watson, J.W. (1970) *North America: Its Countries and Regions*, 2nd edn, London: Longmans.

Wearing, S., Stevenson, D. and Young, T. (2010) *Tourist Cultures: Identity, Place and the Traveller*, London: Sage.

Weed, M. and Bull, C. (2009) *Sports Tourism: Participants, Policy and Providers*, 2nd edn, Oxford: Butterworth Heinemann.

West, S. and Ansell, J. (2010) 'A history of heritage', in West, S., *Understanding Heritage in Practice*, Manchester: Manchester University Press, pp. 7–46.

Westminster Media Forum (2016) 'The future for UK arts, culture and heritage: implementing the new strategy', 7 June.

Weston (2012) 'Sea Odyssey giants give Liverpool businesses a major boost', *Liverpool Echo*, http://www.liverpoolecho.co.uk/news/liverpool-news/sea-odyssey-giants-give-liverpool-3348345, accessed 4 December 2012, 19:00.

Wheeller, B. (2007) 'A stick of rock, cock?: Donald McGill's saucy seaside postcards', in CAUTHE, *Tourism-Past Achievements, Future Challenges*, Sydney: CAUTHE, p. 865.

White, E. (2008) *The Flâneur: A Stroll Through the Paradoxes of Paris*, London: Bloomsbury.

Wiechmann, T. and Pallagst, K.M. (2012) 'Urban shrinkage in Germany and the USA: a comparison of transformation patterns and local strategies', *International Journal of Urban and Regional Research*, 36(2), pp. 261–80.

Wiener, M.J. (1981) *English Culture and the Decline of the Industrial Spirit 1850–1980*, Cambridge: Cambridge University Press.

Wild, L. (2007) *A Year of Family Recipes: Betty's Cookery School*, London: Betty's and Taylors Group.

Wildlife Trusts (2017) 'Some thoughts on rewilding', www.wildlifetrusts.org/rewilding, accessed 26 April 2017.

Wilks, L. and Quinn, B. (2016) 'Linking social capital, cultural capital and heterotopia at the folk festival', *Journal of Comparative Research in Anthropology and Sociology*, 7(1), pp. 23–39.

Williams, A. and Shaw, G. (1997) 'Riding the big dipper: the rise and decline of the British seaside resort in the twentieth century', in Shaw, G. and Williams, A.M. (eds), *The Rise and Fall of British Coastal Resorts*, London: Pinter.

Williams, J. (1992) *A History of Margate's Winter Gardens*, London: Thanet District Council.

Williams, R. (1973) *The Country and the City*, London: Paladin.

Williams, S. (1998) *Tourism Geography*, London: Routledge.

Wills, A. and Phillips, T. (2014) *British Seaside Piers*, Swindon: English Heritage.

Willsher (2016) 'Hi-tech replica to bring prehistoric art of Lascaux within reach', *Guardian*, https://www.theguardian.com/science/2016/dec/03/hi-tech-lascaux-caves-prehistoric-sistine-chapel, accessed 4 January 2017.

Willson, G. and McIntosh, A. (2007) 'Heritage buildings and tourism: an experiential view', *The Journal of Heritage Research*, 2(2), pp. 75–93.

Wilson, R. (1989) 'Changing markets of the English seaside resort', unpublished thesis, Nottingham Business School, Nottingham Polytechnic.

Withey, L. (1997) *Grand Tours and Cook's Tours: A History of Leisure Travel, 1750 to 1915*, London: Aurum Press.

Witmore, C.L. (2006) 'Vision, media, noise and the percolation of time: symmetrical approaches to the mediation of the material world', *Journal of Material Culture*, 11(3), pp. 267–92.

Witmore, C.L. (2007) 'Landscape, time, topology: an archaeological account of the Southern Argolid, Greece', in Hicks, D., McAtackney, L. andFairclough, G. (eds), *Envisioning Landscape: Situations and Standpoints in Archaeology and Heritage*, Walnut Creek, CA: Left Coast, pp. 194–225.

Wöber, K. (1997) 'International city tourism flows', in Mazanec, J.A. (ed.), *International City Tourism*, London: Pinter, pp. 39–53.

Wollheim, R. (1970) *Art and its Objects: An Introduction to Aesthetics*, Cambridge: Cambridge University Press.

Woods, M. (2005) *Rural Geography: Processes, Responses and Experiences in Rural Restructuring*, London: Sage.

www.ccg.leeds.ac.uk (2016) 'Where are the wildest places in Britain', accessed 22 January 2016.

Yagi, C. and Pearce, P. (2016) 'Western castles through Japanese eyes and minds', 6th Biennial ITSA Conference, University of Greenwich London, 19 August.

Yellowbook (2005) *Canterbury: City of Imagination*, Canterbury: Canterbury City Council.

Yeoman, I., Robertson, M., Ali-Knight, J., Drummond, S. and McMahon-Beattie, U. (eds) (2004), *Festival and Events Management: An International Arts and Culture Perspective*, Oxford: Elsevier: Butterworth Heinemann.

York, P. (2016) *The Hipster Handbook*, BBC1, 10 November 2016.

Zelinsky, W. (2001) 'The geographer as voyeur', *Geographical Review*, 91(1–2), pp. 1–8.

Ziller, R. (1990) *Photographing the Self: Methods for Observing Personal Orientations*, London: Sage.

Zukin, S. (1995) 'Whose culture? Whose city?', *The Urban Sociology Reader*, London: Routledge, pp. 281–9.

Zukin, S. (1996) *The Culture of Cities*, Oxford: Blackwell.

Zukin, S. (2010a) *Naked City: The Death and Life of Authentic Urban Places*, Oxford: Oxford University Press.

Zukin, S. (2010b) 'Landscapes of power: from Detroit to Disney World', in Bridge, G. and Watson, S., *The Blackwell City Reader*, New Jersey: John Wiley & Sons, pp. 293–302.

Index